LO ST

MICHELLE PITTMAN

Author: Michelle Pittman

Copyright © 2024

First Published January 2024 Bermingham Books

The moral rights of the author have been asserted.

All rights reserved.

ISBN: 978-0-6459517-8-3

To my family,

We are courageous, strong, and determined.

Our love got us through this, and it will get us through anything.

Thank you for never giving up on us.

Dylan, unknowingly, you gave me the willpower and strength to push us through the toughest ordeal of our lives.

I am proud of who you are, and I wanted to write this book for you so you would always remember that and know you can achieve *anything* in life.

contents

LO ST

MICHELLE PITTMAN

Bermingham BOOKS Where Best Sellers Are Made

October Long Weekend

2017

Chapter 1

GOSFORD WATERFALLS

Seated on the lounge on a spring Thursday night, I turned to Dylan, my 9-year-old, and said, "Let's have a fitness exploring kind of weekend…What do you think?" The expression on his face at the word 'fitness' told me he wasn't too keen. "I have four days off work for the long weekend; how about we plan where to go and visit somewhere different each day?" I said, injecting as much excitement into my voice as I could so he would catch onto my enthusiasm. "We could go bush walking and *explore* different places!"

Dylan loved camping and exploring the wilderness. He had become a massive fan of watching Bear Grylls's survival and wilderness DVDs that had been passed onto him by his brother, Timothy, a huge BG fan also. Once Dylan heard the word EXPLORE, he was all ears!

I explained how I knew of a place I had visited years ago on the Central Coast that people could visit. "You walk through bushland and are suddenly greeted by a beautiful waterfall," I told him. Dylan had never been there before, and I could see in his eyes he was eager to see it for himself. We quickly decided this would be the first place we would visit on our four-day long weekend adventure.

Searching the internet, I showed him some pictures, and the excitement to see it soared. "Tomorrow, we can search where we will go the following day, but we have our first day sorted, so let's get to bed early so we're well-rested," I said, flicking off the computer and standing to get him ready.

When morning came, we both got up early, showered, dressed, and were ready for what was to be the first day of our four-day long weekend of adventures full of exploration. Well, that was the goal!

I wanted to make the most of our four days together since all my other kids were away. My eldest son Daniel, who I commonly referred to as 'My Boy', was twenty-three and had only just returned to NSW from QLD, where he had been working in ocean salvage. He'd been living back at home with us for a while had decided to visit one of his best mates for the long weekend to catch up.

My nineteen-year-old daughter Sarah, or my Hunbee as I called her, had only recently moved out of home a few months prior and was now living in Parkes with her partner. She was madly studying in the retail field and had plans with his family for the weekend.

Timothy, my eighteen-year-old whom I called my Sheeky, was now living in Cessnock in a shared house with friends and working flat out six days a week for a landscaping company. He was the funny one in our family; you could always depend on him for a good laugh.

Smiling while thinking of my other beautiful kids living their best lives, I packed fruit, sandwiches, and bottles of water into a backpack. I picked up my phone that had been charging overnight, along with a spare charger cord for the car so we could keep it charged, contact my older kids if I needed, or they could get a hold of me.

I opened maps on my phone and typed in the address so it could direct us to our first location. I had an idea of where it was, but it had been so many years since I had last been. I wasn't 100 percent sure which street I needed to turn down to lead us to Strickland Falls in Somersby. Location locked in on the phone, I grabbed the car keys. We were ready! "Let's go!" I said excitedly. We were both keen to start exploring.

Arriving at our location, we parked in the Falls car park, got out, and took in our surroundings. Even though it was still quite early, the car park was filling up fast with cars and people. Some with backpacks like me, some with towels draped around their necks. Until this adventure, I had never been a backpacker— seeing other people carrying theirs, I felt like I fit right in. There was a mix of ages mingling around, mainly teenagers, though. We were surrounded by bush, apart from a toilet block, a path that led to some stairs and another track, with a sign saying 'waterfall' and an arrow pointing to the direction to take.

We began walking, following the path towards the waterfall. Sometimes, it had stairs descending; other times, it was flat. We would pass people, some solo, others with friend or family groups, heading back up towards the car park. I guessed they had already finished exploring for the day, yet we were just starting ours.

We followed the path until we came to a large section of open flat rocks. People were sitting on them, kids playing amongst them and splashing in the giant puddles of water on top of the rocks.

Dylan asked, "Is this the waterfall?" looking slightly disappointed.

I smiled and said, "No. This must be a smaller section of it," going on to explain that from memory, the waterfall was so big, you could stand below it and look up to where the water fell from a great height.

We continued down the path. Sometimes, it would become narrow; other times, it would be broader, twisting and turning, until eventually, the trail turned into a dirt track. We smiled and greeted the other people we passed who were coming from the direction we were headed in. At one point, we came to a section where we could see people exploring and climbing over large boulders. Quick as a wink, Dylan asked if we could do it too. I wasn't confident I could climb over them, but I didn't want to discourage him. After all, we were here on an adventure! Searching my face, his lit up when I agreed, saying, "Sure! Let's go!"

We stepped over rocks and big tree logs that had fallen onto the ground. Once we cleared each obstacle, we would resume our journey, following the path until the bush became so dense that we needed to climb over the giant boulders to get to the other side. We weren't alone though; some people were doing the same in the direction we were going as well. Climbing our first whopping boulder, I surprised myself, having doubted that

I could do it. I thought for sure I would never be able to climb over them, but I did!

We kept going and were having so much fun! A few more boulders to climb over, and then we arrived at this big sandstone clamshell-like rock. We stepped gingerly over a small creek with brilliantly clear water streaming over rocks that we used as stepping stones. The sun reflected on the water, which made it glitter. It was so pretty. Enjoying our journey and feeling well accustomed to the boulders and terrain, we continued to climb up, over, and through until we heard our bellies rumble.

"Let's find a spot to sit and eat some lunch, Dylan," I said. People were coming and going, and some walked around the bend of the sandstone rocks. We sat, watching people zigzagging back and forth as we ate our sandwiches, overlooking the creek. We noticed some sections of the path were broader and trickier, with obstructions such as fallen tree branches that you had to either step over or go around.

A man and his friend stopped to say hello to us and commented on the amazing clamshell rock we were sitting on. We agreed! Curiously, I asked them what was beyond the bend they had just walked, and we were yet to. They informed us it ended, and they could not go any further due to thick bush blocking the way. "Oh, OK," I said, "Good to know!"

We each said our goodbyes and wishes to enjoy the day, and then they continued to walk back in the direction we had come from while Dylan and I continued to sit and relax.

We both commented on our surroundings while eating our sandwiches and sitting on the rock in the sun. It was pretty cool how we had climbed over boulders and tree logs to reach this point. We both were quite proud of ourselves!

After finishing our lunch, we drank some water and packed up our remaining food. Once we got to the waterfall, we planned to stop again to eat the fruit we had packed. Hoisting the backpack on, we continued trekking back in the direction we had come from to resume our journey along the path to the waterfall. After climbing back over the boulders, up and over, down and around the rocks and other obstacles, we could finally hear the faint sound of the waterfall. Not yet able to see it, we were excited knowing we were so close now. We passed heaps of people, mostly teenagers, walking along together, laughing and talking with their towels slung over their shoulders or wrapped around their waists.

As we neared closer, the sound of the waterfall became louder and louder. We had arrived! It was such a beautiful sight, with the water cascading down off a massive cliff onto the rocks below. We stepped as close as we could without getting caught in the spray; we hadn't prepared to get wet, and it wasn't exactly a hot day either. The sun was shining, but it wasn't beaming like it would on a summer's day. It didn't worry the teenagers though! We smiled and enjoyed watching them having fun.

There was a large, flat rock behind the waterfall where people could stand and then step forward, moving in and out of the waterfall. Looking down towards the bottom of the Falls, people were also standing underneath it. We could hear some saying to each other how cold the water felt. Others, like us, were happy not going in the water, opting just to enjoy watching everyone else's reactions to the cold water.

We took some photos of the waterfall as we talked about the large volume of water that constantly dropped from the cliff. It amazed us—it never seemed to end! After a while, we decided to head back up the path to another section we had seen on the way down. With one last look at the Falls, we turned with smiles on our faces and talked animatedly about what we had seen as we began heading back up to the spot conveniently on our way back to the car.

We had been walking and exploring for hours, so we decided to stop at a section that overlooked a large part of the surrounding bush. We stood and ate some of the fruit we had brought with us. With the food and drinks nearly gone, it made the backpack so much lighter and easier to carry. We could feel we were beginning to get tired—it had been a big day out. Deciding to call it a day, we returned to the car for the drive home.

Reaching the car, we talked about how surprised we both felt, conquering those whopping boulders and the giant tree trunks across our path. Feeling accomplished, we arrived home, showered, ate dinner, and relaxed on the lounge to watch another episode of Bear Grylls. We needed to decide where we were going the next day, too, to plan ahead.

Searching the internet together on my phone, we found a place called the Yengo National Park. Not only had we never been there before, we had never heard of it either! It described walking tracks with plenty to explore. It sounded great to us! With both of us happy with our discovery, it was agreed, this would be the place we'd explore tomorrow. I got up and, for no particular reason, grabbed an A4 piece of paper to start documenting the places we had visited or planned on visiting. I placed the pen and paper on the bench, turned off the TV and lights, and both went to bed.

Chapter 2

YENGO NATIONAL PARK

I always woke up earlier than Dylan, and today wasn't an exception. I bounced out of bed excited and keen to continue our adventures. I had my morning shower, got dressed, and began preparing the backpack. I thought I'd pack four water bottles for this trek and was putting them in the pack when Dylan woke up. He had his shower and got ready for the day, and we both had breakfast together, discussing what food we would pack for this adventure.

We decided on similar items to yesterday—sandwiches with a different filling, peanut butter and lettuce, which was one of our favourites—the bottled water, and this time, I would also cut up a container full of watermelon. I loaded the items into the backpack and picked it up. It was heavy, but remembering from the day before, I knew it wouldn't be for too long, just until we began to eat and drink some of the food and water. "Let's go," I said, picking up the mobile phone and backpack. Dylan didn't have to be told twice; he was already heading out the door!

During our drive, Dylan was on my phone searching for what we would see at the Yengo and relaying information about the different trails we could go on and the bushland there. Dylan had always been good at reading and understanding directions, so while I drove, he gave the directions. He could always remember previous directions and places he had been to. He visited Queensland with his older brother, Daniel, for the first time when we drove up for a visit. Two years later, Dylan remembered which exit to take, the lights to turn at—everything, all the way to Daniel's house! I was amazed at how he could remember it all from two years prior.

On the other hand, well, when we go to the shopping centre, I often can't remember where I've even parked my car! Dylan remembers precisely where the vehicle is parked and which direction, level, and aisle we left it! He expertly navigated us today, too, leading us effortlessly to the entry of the Yengo. We drove along the dirt road, winding up the hill and around the bends. The National Park was surrounded by thick bushes on either side of the road. When we arrived at a cleared section, we saw other vehicles parked, so we decided it looked like a good place for us to park, too.

Getting out, we noticed every direction was bushland. Putting the backpack on, we locked the car, and I placed the keys in the small side pocket of the pack. A high fence line ran along the car park's left side. We walked over to it and saw a big drop down to the unknown beyond the fence, among a sea of greenery.

The grassy area near the car park had some BBQs and a small section to camp. Seeing it, Dylan's eyes lit up. He was keen and asked if we could camp here with Daniel one day. My eldest son is a huge camper. His ability to light and maintain a campfire is legendary in our family.

Walking along the grass, we passed a few people saying hello to us before coming to a section where the road went in different directions. A sign announcing a walking track caught our eye. Looking in different directions, we could see nothing but green trees and dense bushland, some sections heading down, while other sections were heading up high.

We stood for a while, thinking, reading the signs, and then walked along the road to the sign that said 'walking track'. At the beginning of the track, off to the right, the road ahead went down a steep hill to where two thick poles sat on either side of the road linked by a low hanging chain about knee height from the ground. Across the road, a sign said it was for emergency vehicles only. We came to the beginning of the track and began walking. It was fairly narrow, so Dylan had to walk behind me.

After some time, we came to an intersection. One direction had a sign blocking the way that said, 'No Entry' due to hazardous conditions. We were curious about what the hazard was. Between the two of us, we took turns guessing what it could be—maybe trees had fallen down, blocking the way through, or perhaps part of the track had collapsed.

Curiosity didn't win out. Playing it safe, we continued on the track without the sign, which soon became narrow. I was keeping an eye on the left side of us—it was a loooonnnng way down. I was unable to see the bottom, and I wasn't willing to stand close enough to the edge to find out either! I got an uncomfortable, eerie feeling in the pit of my stomach. I carefully turned my head to tell Dylan to stay as close to the right of the path as possible. The trail snaked around bends, twisting, turning downhill, then up.

We came to a set of stairs. Not concrete stairs or anything; instead, they were pressed dirt blocks with logs dividing each step down to where we arrived at a small, narrow bridge. It was made from wooden planks wedged close together. Standing on the bridge looking down, we could see how uneven the ground was and that it was covered in different-sized tree branches, weeds, and shrubs growing both on the ground and over the tree branches.

The bridge had no side rails, but the planks were wide enough to walk across safely. We cautiously remained in the middle of the planks as we walked across them. With each step we made, the wood would groan and creak. We began to joke and laugh, saying it sounded like a troll was underneath the bridge and wanted to know why we were crossing it. Laughing, we pretended to reply to him, saying, "Mind your own business!" Ten or so more steps, and we were off the bridge and back on the track.

The bush was so thick we felt small and engulfed by it. I could see the track ahead was coming to a left-hand bend, and we wouldn't be able to see what was further on until we reached that point. As we walked, I peered down. We were getting higher and higher, which freaked me out a little. I'm a bit scared of heights. I told Dylan, "I don't like this. The track is getting narrow, and we are getting closer and closer to the edge."

Dylan, being Dylan, said, "It's fine, Mum. You're just scaring yourself." His confidence settled me. Maybe he was right?

We continued walking, the track snaking continually—up, down, left, right, around until we came to another section where the ground to the right was a bit flatter—maybe the size of a picnic rug. We decided to stop and sit while we had the opportunity. I felt relieved to be taking a break on this side because on the other side was the big drop. It felt good to remove the weight of the backpack. I shrugged and rolled my shoulders to stretch them out.

Sitting on the ground, I took out our sandwiches and two water bottles. We sat facing the drop side while eating and drinking. Looking as far as we could see, we noticed how large the area was. Covered in millions of trees and bushland, the view was spectacular. We spotted cleared little regions of the landscape and pretended what our house would look like if we were to build one and live there.

With our sandwiches finished, we started on the water-melon, talking and taking in the view. It was so breathtaking we didn't want to leave. "Let's build a house right here. It would be a cute little cabin," I joked. Dylan was laughing and visualising our little cabin in the bush, saying how we would place a table and chairs on our veranda, sit outside, and eat breakfast every morning before work and school.

Packing up our lunch rubbish and containers, I put them in the backpack, got up, hoisted the pack into place over my shoulders, and we continued on after taking in one last look. Following the track, the right side where we walked became walls of rocks; the left side was still the significant drop and open to the views. While walking through this section, I advised Dylan that we should use the rock as support as the track had begun leading uphill again.

I checked my phone and saw it had been a few hours since we began our trek from the car park. I also noticed we had no reception. We arrived at the top of the hill and then came to another section that was like a flat picnic area. We stopped to check the view and noticed we were much higher than our last stop; so far up, we couldn't see the bottom at all. Looking directly out from where we were standing, I could see a long way into the distance. Between us and the farthest point, all you could see was the top of the treetops spread out like a green canopy.

I took another bottle out, and we drank its contents. By this time, we had drunk two bottles—one remained. I was happy the bag was getting much lighter—thank goodness! My shoulders had begun hurting, so lessening the load was a welcome relief. I suggested we finish off the watermelon while we were standing still and taking a rest, to which Dylan readily agreed. Hungry, we virtually inhaled the fruit. Minutes later, I was snapping on the lid of the finished container and stowing it away again in the backpack with the three empty water bottles before we began on our way again.

The side where the rock had been offering some support had now turned into bush again, the track still going uphill on an incredible angle. Continuing on a while further, I wondered how long it took to walk the track, and where we would end up-since we were still hiking uphill. I was talking to Dylan about it and looking at my phone to try and calculate it. There had been nothing on any of the signs we had seen. We had walked for hours, and the end didn't appear to be in sight. I thought it might be best to turn around and return the way we came. Dylan agreed, saying he felt it was enough for the day. "OK, let's turn around and head back then."

There wasn't a lot of room to turn around, so I told Dylan to step back as far as he could into the bush side, so I could step around him and be in front again to lead us down. That way, if I slipped or lost my balance, he would know to be more careful or wait for me to help him. Forever laughing and joking together, we made fun of that, too. I told him, "If I happen to fall off the edge, definitely don't follow me, OK?" We laughed some more as we navigated the switch around and began walking downhill. With the track being dirt littered with a lot of rocks, we found our footing did take a slight slide here and there, but not enough for us to slip over or fall. It was just enough to feel the rocks sliding between our shoes and the ground.

Walking back, we remembered the familiar sections or places we stopped at or noticed on our way up, which became our benchmark of how close we were to getting back to the car. We stopped at the section where we had eaten our lunch to take in the views one last time and sip some more water before we kept walking. Standing there admiring the views and catching our breath, I reminded myself and Dylan that we were not familiar with our surroundings, nor did we have any hiking experience, so we had to be careful to keep following the trail.

Downhill and uphill, we followed the trail to the bridge we had previously walked across. Immediately, we began joking about the troll again until we reached the other side. Walking on, our jokes sometimes would run their course, overstretch, or be taken too far, but they were our jokes that kept us amused, made us laugh and seemed to make the walk back to the car far quicker. We were moving our asses, knowing we would soon be reaching the car park.

Once there, I noticed there was not one other car—just ours

stood lonely among the bush. No campers, no picnickers, no walkers, no one. I commented to Dylan that we made the right decision to turn around when we did.

I dug my phone out of the backpack and saw only a few hours were left before sundown—we *definitely* made the right decision to stop and turn around!

Getting into the car, we both felt relieved to sit down. I dug the remaining bottle of water out of the pack, placed it in a cupholder, and retrieved the last of the fruit—two apples and two bananas to eat on the way home. I plugged the phone charger in, got the car keys, started the ignition, and we were off back to the safety of home!

Dylan asked, "Do you remember the way back?"

I laughed. "I think so," I replied, smiling.

He laughed, knowing fully well how hopeless with directions I was, and assured me he remembered the way. Having complete confidence that he would remember the way back, I didn't use the navigator, and sure enough, he guided us expertly back to Singleton.

We were exhausted from the day's adventure. "Let's shower, have something quick and easy for dinner, sit on the lounge, and search where we want to go tomorrow," I suggested.

He agreed.

Later, sitting on the lounge searching the internet, we came across a place called Walka Water Works. I'd never heard of the place. Intrigued, we found out it was only a short drive from Maitland, a place familiar to both of us. From there, we could use the navigator to help Dylan direct me from Maitland to the Walka Water Works place. Putting the phone down, we both got comfortable on the lounge, watching Bear Grylls. Tonight was an episode of Man vs Wild, where Bear demonstrated different strategies for survival using nature.

I woke feeling cold; we both had fallen asleep on the lounge. Standing up, I gently woke Dylan and said, "Hop into bed." He was barely lucid as I held his arm to guide and support him to his bedroom. I tucked him into bed and said, "Goodnight. I love you."

No reply from him. He had fallen right back to sleep. I left his bedroom, turned off the TV in the lounge room, walked into my bedroom, and got into bed. Typically, I turn on my TV to watch to help me fall asleep, but tonight, there was no need. I felt so tired and wrung out from all our walking and exploring, and within a few minutes, I was asleep.

Chapter 3

MAITLAND WALKA WATER WORKS

I woke to see daylight. The sun was shining through a slight curve I had in my blinds. Stretching and getting out of bed, I walked over to my window, moving one of the blinds across to see the sky. It was bright blue, not a cloud in the sky, and I could feel the sun was already warm on my skin. "What a beautiful morning," I thought, picking up my phone to see the time. It was 8 a.m. "Wow! I have slept in a lot!"

My routine is usually that I'm up at 5:30 a.m., even on my day off, and I leave home to arrive at work for a 7 a.m. start. I'd slept beyond even that.

Dylan wasn't awake yet, either. On the mornings I am still home; he wakes up and comes into my room to say good morning. Keen to get a start on our third adventure day, I showered, got dressed, put on some socks and shoes, and tied my hair up, ready for when Dylan woke up. I sat on the front veranda, taking in the morning sun, drinking my cup of tea, which I like to call my 'go-go juice', as I did every morning. Halfway through it, I heard Dylan getting up and using the bathroom.

He came out and sat beside me on a chair, and we both asked, "How did you sleep?"

"So good!" we both replied to each other. I asked him if he remembered me waking him up from the lounge watching BG, and he couldn't.

I said, "You must have been tired!" and we both had a little laugh.

Sitting in the sun, Dylan asked, "You're all ready to go, Mum? Are we still going to the place at Maitland?"

I stood up, picked up my cup, and said, "Yes, we sure are!" as we walked back into the house. "Have your shower, get dressed, and we will head off."

After Dylan had his shower, got dressed, and was putting on his socks and shoes, he asked for his breakfast. Usually, I didn't eat a proper breakfast. My breakfast consisted of a cup of tea, but I always prepared his.

While he ate his breakfast, I began preparing our lunch and fruit. Again, I packed four 600ml water bottles for today's adventure. Dylan was happy to have the same as yesterday except for the fruit. Today, he opted for straw-berries, pears, and grapes. We always have a variety of fruit in the house. I take different fruit to work every day for lunch, plus a water bottle or two, and he usually does the same. This time, I also packed two muesli bars, remembering that we became hungry often while exploring on our adventures.

Once again, I could feel the weight was heavy, but the excitement we both felt to be heading off again overrode it quickly. Collecting my keys and phone, I swung the backpack over my shoulder, and we headed out the door to the car. I was proud that we were accomplishing our goals. My mum always used to tell me that I am a person who usually does what I say I will.

Arriving in Maitland, Dylan was ready with the navigator and gave me directions to Walka—turn next left, make a right up ahead. Passing the hospital I had driven past many times before this day. I commented how I had never even noticed the sign saying Walka Water Works, and Dylan agreed.

Driving along, we could see farmland and horses in the distance and a tall smokestack with no smoke rising out of it. It was made of bricks and stood tall above the crops. That must be the place, we thought. Arriving at the entrance gate, we could see the tall smokestack we spotted on the drive in, which was even more significant in height up close.

We decided on where to park. In front of us was a large green grass area where many families had set up ball games and picnics. Kids and adults alike had their bikes and were riding around. I pointed to a pond that curved around a small knee-high brick wall built around a section of the pond. I got our pack from the back seat, put it on my back, and locked the car.

Behind us were different-sized brick buildings. Dylan asked what they were. I said I wasn't really sure, but suggested we head over and have a look. Walking to the first building, we couldn't go inside as the big doors were locked. We peeked inside the window panel on the door and saw some chairs scattered around a rectangle-shaped furniture piece that looked like it may have been used as an old counter. The furniture was old and like nothing we had seen before or were familiar with. From what we could see, we guessed the building had been used as a cafe or something similar. We had no clue, but we often liked to guess, use our imaginations and create a story about things we saw to share with each other.

Walking to the next building, we looked up and could see the smokestack. It was so tall! It didn't even look that tall from the car park! We were able to walk around and see inside some of the buildings next to it. Again, we could see some old furnishings that I thought could be antique and worth a lot of money, from what I had learnt watching antique-style programs and hearing from different people about old furniture. Even though this furniture wasn't in good condition, I figured it was still part of the history, so it was probably worth something.

We were curious about what this place used to be. As we walked around it, we noticed signs and posts with information. I read them aloud to Dylan so he also knew the history of the place. From the information provided, we learnt it was a 19th-century pumping station, which surprised us. We thought it may have been an old school or café, a place of worship—who knew, clearly not us, but we had some fun and laughs trying to guess!

We walked onwards to a large open grass area, where a round fenced-off section dropped down into another grass area with some shrubs. The fence was high—over 8 feet! We again wondered what it was used for. "Maybe a pool?" we laughed, running away with a wild story as always.

It was the proper depth for a pool, and under state laws, pools need to be fenced like this. Maybe it had been filled in? Hence, the grass and shrubs that resided there now. It was a long way down, and if anyone fell in it, they would find it difficult to get back out. It seemed like a logical choice to us!

Looking in the direction we were going to walk next, we heard a sound. Quickly scanning to see what it was, we suddenly saw a tiny little yellow train. It was so cute! It wasn't too far from where I had parked the car, but because we had decided to walk in the opposite direction, we didn't notice there was a small train station as well. It was like something you would find in the kids' section of an amusement park or a carnival.

We could see people sitting on it, including adults. This confirmed to us we could go on it too! As we got closer, we could see the miniature station and the train tracks leading off into the bush and disappearing from sight. Excited and keen to take a ride, we joined the queue, where some kids jumped up and down, excitedly waiting for their turn.

The driver introduced himself and asked all passengers wishing to ride the train to leave any bags inside the gate but bring any valuable belongings. I decided to keep my keys, wallet, and mobile phone on me. The cute little station looked old and in need of some TLC. If we hadn't already seen the train in motion, we wouldn't have believed it was in use and would have thought it was part of history only. It amazed me, and we were both excited to be lucky enough to have a turn on it.

Dylan said with a big smile on his face, "Imagine having this in our backyard, Mum?" We laughed.

I said, "You'd never get off!"

Dylan agreed, still laughing.

When our turn came, we were instructed to sit on the little train as if riding a horse, with one foot on the left side of the long cushion beam and the other on the right side of the seat.

We sat as instructed and waited for the other passengers to be seated. The driver then instructed us not to extend or reach out with our arms or hands while the train was in motion, but we were welcome to take photos if we wished. With one last instruction to keep a firm hold of young children, to which Dylan, sitting in front of me, turned his head to look back at me and smile, we began to roll forward.

The little train went far into the bush, and we could see many sections, with the driver explaining the different parts and what each was called and used for. We went across small bridges; one looked like a pint-sized replica of the harbour bridge. It was so cute! We went over creeks and stopped at a place where other engines and seating trailers were sitting with silver shed-like structures. A few workmen were talking and walking around. Spotting us, they waved, and we all waved back.

Rolling on, we passed people walking or riding their bikes, stopping at the crossing section for the train to go through. Looking at the ground, I could tell they were following a walking track. I leaned forward and said to Dylan, "We are definitely walking that track when we get back." He turned to me and smiled in agreement. As the train rolled on, Dylan often leaned back towards me to speak, commenting on different things, and I leaned forward to do the same.

The train ride took us through a large section of the surrounding bush. We were so happy to be lucky enough to get a turn, we were beaming on the return run. As we pulled up to the station, more people had lined up to have a go as well. It was getting busy! Judging from the now packed carpark, with even more carloads of people still arriving, we had come at the perfect time. "This is such a popular place!" I said to Dylan.

"Could we have something to eat?" he asked.

I replied, "Sure, we sure can! Where would you like to eat?"

He suggested going over to the grass where the play park section was.

I looked over to where he was pointing and saw a swing set that parents and other kids were playing on. Kids and adults were kicking balls, playing cricket, sitting in their fold-out chairs, and eating and enjoying the sun on their day out. There was activity and people everywhere, with very little space to sit—kids, and adults were everywhere, with balls flying through the air in all directions. It looked hectic!

I turned to Dylan and said, "What if we find the entry to that path that the people we saw earlier were walking on? Then we could see if there is somewhere there we can stop and eat?" I added, "We will get hit by a ball or stamped on by all the kids and adults running around sitting there!" and laughed.

Dylan agreed, "Yeah, there are heaps of people here, Mum," he said, walking in the direction we believed would lead us to the path's entry.

On our way, we saw a sign confirming we were headed in the right direction to the walking track. It was a wide, concreted path. On one side was the pond with the brick wall. Off in the distance on the other side were a few scattered houses, with long grass between us and the homes. Looking at the brick wall, I saw it wasn't too high, so I thought we could sit on it while we ate. Pointing at it, I said, "Let's go over to the wall and eat there."

Sitting on the edge of the wall overlooking the pond, we watched ducks and birds gliding across the water. While eating our lunch, we could see and hear the families over at the large grass park picnic area. Some people walked past us on the track;

others were riding their bikes. Dylan said, "I should bring my bike here to ride around."

I replied, "Yeah, next time we come here, we will bring your bike, definitely."

When we finished eating, we took a last sip of water, packed everything into the backpack, and began walking the path. Walking side by side, we laughed at the antics of the ducks and birds splashing and playing in the water.

Dylan said, "Nan would try and catch them!"

My mum, Teresa, being an animal lover, loved ducks. She would keep them as pets if she could.

We walked as the ducks waddled back and forth into the water and out of the water.

The concrete path turned into a bush track, and we were officially on the track. We could see people up ahead, and sometimes, we would hear voices coming up behind us as well. We would move to the side as they neared to let them pass. Some walked, some were on bikes, some were adults, and some were families with kids of all ages. The little kids were so cute on their bikes, wearing their helmets, and trying to ride through the bush.

Sometimes, we would see and hear the train coming through the bush, and a little further along, we saw it emerge clearly from the bush. We smiled and waved at the passengers, hoping they felt the same as we did while on the train. They smiled and waved back, so we guessed they were as happy as we were to get the chance to ride the little train.

As we walked, we sometimes needed to cross over the train tracks, which were marked with signs saying, 'Please Look Before Crossing For The Approaching Train'.

We kept following the track in the direction it led us, and after a while, we came to a crossroad with multiple signs explaining the different tracks we could choose to follow. Unsure of which one to walk, I decided to take the track to the left, remembering Mum's golden rule of advice—if unsure or stuck on which direction to go, always choose left.

The dirt track was still wide, so we walked side by side. We didn't see any people while walking and weren't sure where this track would lead us, but we were confident it would lead us back somewhere near the carpark. I removed one arm from the backpack strap and pulled it in front of my chest as we walked to get the remaining fruit we had out of the bag, which had become much lighter than it was when I first hoisted it into position on my back at the car.

Finishing off the fruit and drinking water from our water bottles while walking, we could hear people's voices and cheering in the distance. Even though we couldn't see anyone, we could tell they were playing some sort of game. We took turns guessing what game we thought they were playing—cricket, football, soccer, tips, or hide and seek.

We continued walking, going around a right-hand side bend. The track straightened out, and we saw some houses on our left. They were far apart and on acres of land. I always wondered what people did when they lived on large parcels of land out in the middle of nowhere. I assumed they would drive to their jobs, and when home, they would be busy looking after their property.

We continued and came to an intersection. Left or right? I shared my thought process with Dylan, explaining that I thought if we went right, it might take us back around to where we began, and from there, we had some other track choices we could take if we wanted to. Remembering my mum's number one rule when it came to uncertain direction, with no further questioning or doubt, we quickly chose the left option. "OK, let's go this way then," I said, pointing and walking towards the left.

We could still hear lots of people in the distance while we walked and pointed out different things to each other. It wasn't too long before we again arrived at another intersection. To the right was a gated dead end with a sign that said, 'Authorised Vehicles Only'. While standing there, we saw cars driving up to the gate before turning around and heading back in the direction they had come, which was the same direction we had been heading. Left again, we thought.

This time, the track didn't arrive at an intersection. Instead, it opened up to the green grassy area with the high fence and 'pool' we had seen upon our arrival. People were playing a game of cricket, and we wondered if that was the game we could hear when walking on the track. Up the hill, the brick smokestack towered.

I asked Dylan if he had had enough for the day. I knew I did! I was buggered and hoping he would say yes.

"Yes, my feet and legs are tired," he said.

Mine felt the same. "How about we buy Chinese for dinner tonight?" I asked.

We both loved Chinese food, and Dylan excitedly confirmed he was happy with my idea. "We'll go home, have showers, get dressed, and head back out for Chinese then?"

As we drove, we recounted all the fascinating things we learnt about the Walka Water Works and how surprising it was to see so many people there enjoying their day out. Feeling very pleased and happy with ourselves, having done another day of exploring and adventuring in a place we didn't even know existed, we arrived home, unpacked, and I said to Dylan, "You go have your shower first, Possum," calling him by the nickname I had given him as a baby after he began to crawl. One minute, he would be right at my feet; the next minute, he had taken off down the hallway as quickly as a possum.

While waiting for him to finish his shower, I fed and topped up Bunny Girl's water. She was a little black rabbit who lived in our backyard and was very much part of our family. Her original name was Angel, but I began to call her a criminal after she continuously dug in the dirt under her cage to escape and run around free. When we first got her, she was free and not caged, but after a cat tried to get her, and with dogs around us, for her own safety, we bought her a hutch to be housed in. She wasn't a fan and would try all sorts of ways to escape. Quickly realising she was no Angel, we began calling her Bunny Girl, which she's been called ever since.

I could hear Dylan out of the shower, so I headed to have mine. With both of us refreshed and ready, we headed out for Chinese. We already knew what we were ordering before we arrived. Our favourite—two short soups, one large special fried rice, and one combination of sweet and sour and honey king prawns. Usually, we always ask for our leftovers to be put into takeaway containers for the next day. I always thought, why not for the price we pay, and it's so good, you wouldn't want to waste it.

Sitting at the set of traffic lights that led us home, Dylan pointed to a sign across the road and asked what it said. The sign was large and brown with white writing. "It says, 'Mount Royal

National Park'," I told him, surprised, noticing the arrow pointing in the direction we were turning. In all the times I had sat at or passed through those lights, I had never noticed or paid attention to this sign. I was shocked I hadn't seen it before!

I told Dylan how amazing it was that there's a national park right where we live. "Tomorrow, we will go check it out! It's a good thing you noticed it!" Being just down the road, I thought we wouldn't even have to drive that far, still amazed how I had never noticed the sign before.

Pulling into our driveway, I could feel my muscles and body aching from all our walking and movement over the past few days. It felt like a real effort just to get out of the car. With our Chinese leftovers in the fridge, we got into our pyjamas and laid on the lounge to watch another episode of Bear Grylls's show before bed.

Chapter 4

MOUNT ROYAL NATIONAL PARK

Day 1 MONDAY

It was the last day of our October long weekend adventures, which meant it was back to work for me tomorrow.

We had a slow start to the morning, still feeling exhausted from the last few days of adventuring national parks, waterfalls, climbing boulders, rocks, and tree logs, walking up hills, down hills, around cliffs and ponds, as well as the packing, driving, unpacking, and whatnot each day.

It was 9 a.m., and Dylan was still asleep. Outside, it looked overcast with clouds forming. I thought there could be a chance of rain, and it probably wasn't a great idea to go bush trail walking today. After visiting the Yengo State Forest, Gosford Waterfalls, and the Walka Water Works on clear sunny days, I thought we shouldn't push it, so I let Dylan sleep in and was thankful for the time to rest in bed as well.

He woke around 10:30 a.m. with a big smile and keen for breakfast. I prepared some mangos and juice for us both. The mangos were the big, juicy, kind…and delicious! We loved eating fresh mango. I raised with Dylan that due to the overcast conditions, we would not go bushwalking, but could still go for a drive and check the Mount Royal National Park out for a future trek as he'd been excited to discover the sign the night before.

We were curious; after all, it was so close to home—it seemed like it was meant to be. Only two streets away from home, sitting at the intersection to turn right, I noticed the same brown sign with white writing and an arrow. I couldn't understand how I had ever missed it. I would stop at this intersection six mornings and six evenings a week on my way to or from work. How had I not seen it before?

Leaving home, I told Dylan I wouldn't pack food or bring multiple water bottles or even a backpack as we had on our previous days since we weren't planning on walking a long way. Instead, we would take only a water bottle for each of us as we would typically do when heading off anywhere. We wouldn't be out all day, so I wore a black cotton T-shirt with 3-quarter denim jeans, cotton socks, and laced-up joggers. Dylan wore black cotton shorts, a red cotton T-shirt, socks, and joggers.

Today was going to be an easy, relaxed kind of day. We weren't focused on exploring or getting the exercise like we had been; we were just taking a curious drive and a little walk. In a way, I think we both felt relieved and happy about it. I think that was why we initially got moving to get ready and jump in the car in the first place—because it was going to be an easy day.

After we had both grabbed a bottle of water from the fridge, I picked up the car keys, my wallet, and my mobile phone and was a little excited to find our discovered place from the night before. Getting into the car, we buckled up and off we went. It was a bit of a drive before we reached the entrance to the Mount Royal, much longer than I had anticipated. The road, however, had luckily been smooth as it was tarred the whole way. The countryside was so green, pretty, and quiet.

At one part of the journey, we both noticed a small object in the middle of the road. Dylan said, "What was that, Mum? What is that?"

I had no clue and said, "I'm not sure." I slowed down, and as we got closer, we could see it was a small animal moving at a very slow pace. I was not close enough to make out what it was, but my first thought, going on my experience travelling on many country roads, was that it could be a small turtle. I had stopped on several occasions to or from home in the middle of the road, exited my vehicle, and carefully picked up a turtle to carry it to the side of the road where it would be safer.

We are both animal lovers, in fact, our whole family is. I slowly inched the vehicle forward, carefully approaching the small animal so we didn't scare or startle it, or worse, run over it. Luckily, we had no one behind us impatient to get around us. Not many cars seemed to travel on this road. It was mainly residents who lived in the houses situated on the acres and acres of land on either side than anyone else.

Very close to the animal now, we were surprised to see it was an echidna trying to cross the road. We stopped completely and watched the little one ever so slowly cross the road to safety. I kept a check in the rear-view mirror and up ahead, ready to alert and warn any vehicles coming along of the echidna's slow trek. Dylan was surprised at all the spikes and wondered if they were sharp. To be honest, I didn't really know, but I assumed they would be given they had them for protection from other animals. I shared my thoughts and added, "Thank goodness he made it to the other side, because there was no way I was going to pick him up!"

Dylan laughed and agreed.

Dylan picked up my phone and reached out his window to take a photo once it had reached his side of the car. I leaned across to his window for a closer and clearer view. We wished the little guy good luck on his adventure before he ambled into the long grass and disappeared on Dylan's side of the road. "Wow! That was so cool to see," I said.

Dylan agreed. "Let's get going. We may see some more animals like crocodiles or a T-Rex dinosaur," we joked.

Hitting the accelerator, we were on our way, feeling happy that not only did we do a good dead looking out for our spikey friend, but we also got to admire him as well. We had never seen an echidna outside of a zoo, so it was a cool experience seeing one in the wild. Driving with caution and on the lookout for more echidnas or other animals we might see, we spotted some horses playing and resting in paddocks, cows, and birds. A trip along a country road never disappointed—there was always something to see and appreciate.

Arriving at the entrance to the Mount Royal National Park, I came to a complete stop, noticing the driving track leading into the park was rugged and covered with small rocks, small branches, and leaves. We cruised cautiously ahead on the unfamiliar terrain that steadily declined.

As we drove slowly along, getting deeper into the belly of the park, I kept having this eerie feeling that something was going to happen or go wrong. It was like my gut was telling me something was going to happen. It didn't feel like something serious or really bad, it was more like we might get a flat tire due to the rocks we were driving on. We could hear them underneath the tires, and some made a very disturbing sound under the wheels and the car. Not good, let me tell you, but having a spare tyre and a car jack in the car gave me some reassurance that we would be okay if something did happen.

Dylan, being the young, carefree child he was, didn't give it a second thought. He was enjoying it—I could see. I did the worrying like mums do and hoped that if we did get a flat it would only be one, because I only had one spare. Any more flats, and we'd be fucked! I tried to settle myself. "Now you're being paranoid, Michelle! Calm down," I said under my breath, as I stole another quick glance at Dylan, who was now taking photos of the bushland.

From here, we could see we were still quite high up, even though we had driven down a fair way already. The views were amazing and worth the nervousness I always felt when it came to heights. Still feeling uneasy, my bad gut feeling that said something was going to happen hadn't left me. I just didn't feel confident here. Maybe it was the flat tyre scenario still worrying me.

I thought I could change a tyre if I needed to. Even though I had never needed to, I'd seen plenty of people who had and had an idea of where to position a jack, how to undo the bolts, remove the rim, place the rim on a tyre, and whatnot. I hoped I knew enough. Not seeing anything else that gave me concern or any signs of immediate danger, I continued to drive further in and wrote the gut feeling off as just normal nervousness.

We heard the motorbike before we saw it heading towards us. I was already going slow, but slowed down even more so we could pass each other as the trail road was quite narrow in parts, and not wide enough for two car vehicles to pass each other without one having to move right over. To be extra careful, I stopped while the motorbike passed. I noticed it was a male rider, not that it made any difference male or female, but I noticed. To be clear, I know nothing about motorbikes except they have two wheels, handlebars, and are loud! But I knew enough to know it wasn't fancy like a Harley or anything— it looked like a regular trail bike to me.

The bike was heading out of the park. "Well, if he can do it on a motorbike, surely we can in a 4WD!" I said, smiling at Dylan. He nodded, confidently agreeing, so I continued slowly, always keeping one foot on the brake, especially when the track became steep and even more windy. I felt prepared, going more than slow enough that if I needed to stop quickly for anything unexpected like an animal darting in front of us, or another vehicle coming towards us, I could.

Rocks crunched under the tyres, sticks snapped, and the bed of leaves ruffled as we drove further and further down until we finally arrived at what I'd loosely call a camping ground. We were now at the bottom of the park, I thought. Trees towered above us and seemed to loom into the heavens, perching from the cliffs above.

"We started all the way up there," I thought aloud, staring up at the cliff edge, following the trail road with my eyes down to where we were now idling, checking the grounds out. We decided to park the car and have a look around. I said to Dylan, "Leave the water bottles. We won't be gone long."

We walked around, curiously exploring our surroundings— keen to see what we had just driven into. It was like some hidden, long-forgotten, deserted place we had happened to stumble across. I noticed the grey clouds were still looming in the sky through the trees, but so far, no raindrops. I felt we had a little time before the rain came to look around quickly.

The camping ground was huge, with a large fire pit and wooden outdoor seating. Dylan and I, at the same time, thought the same thing—Daniel would LOVE that fire pit! He would be in his camping glory being here. We took multiple photos so we could send them to him. That's when I noticed there was no phone reception. I guess I shouldn't have been too shocked, given that we were so far down in the national park, but being so close to town, I was slightly surprised.

"Oh well," I thought without a care, we'd just send the pics to Daniel when we got home. No problem. We continued to walk around and saw a noticeboard. I began to read it aloud to Dylan. The board named and pictured all the different trees in Mount Royal National Park, along with the birds that lived here. There was also a sign sticking out of the ground pointing to where the toilets were and where the trail started. I knew we would have to check that out.

We had no expectations of wandering far enough that we'd lose sight of the car or even need the water bottles. We intended to see the toilets and where the track started, then return to the car. We walked along the grassy clearing and followed a tiny walking track that led to an old wooden shack. A sign pointing in its direction announced it as the toilet block. We walked closer to take a look. It was old, and the wooden door gave a big creaking sound when I pushed it open.

I took one look inside, "It looks scary," I said, feeling the hairs on my neck stand up.

"Like a horror movie," he said, and we both laughed. My mind was racing. The toilet was tucked off to the side of the track, engulfed by bush. It truly was like something straight out of Friday the 13th.

"If we need to go, we're going in the bush!" We laughed nervously and began walking away to see where the trail started; eager to put some distance between us and the spooky toilet.

The trail bent around, and feeling a few raindrops on my face, I looked up, seeing the sky becoming even more grey and cloudy. "I think we should head back in a few minutes before it really starts to rain," I said. Dylan nodded in reply. We walked around the bend, which was steep and covered in rocks, leaves, and sticks, just like the road in was. The walking path was so thick, we couldn't even see anything in the distance, just miles of bushland and mountains covered in bushland. It was intriguing, this place.

I really should have listened to my gut and called it a day. I should have to turn back there and then, but I didn't. I should have considered Dylan's feelings. He had already indicated he was keen to leave too, but instead, the stupid psychological bullshit I fill my head with took over. Instead of listening to my better judgement and gut feelings, I pushed my inner voice aside and continued to encourage both Dylan and myself on by saying, "Well, we've started the track now. We should just finish it and complete it." Trust me as I say this: I, too, am hitting my palm on my head writing this.

On we walked. As we reached the end of the track, it was like walking on the edge of a dried-up creek. It was completely covered in small grey rocks. We stood talking about them for a short while and taking in what we could hear and see, which wasn't much since the bushland completely surrounded us. I could feel the temperature dropping and knew without looking at my phone, it was getting on in the afternoon. I wasn't sure how long it had been since we left the car, but at a guess, I would say maybe 3 or 4 hours.

Looking beyond the dried-up creek to the other side, it looked like the walking track continued, so I suggested we walk straight across to it, have a quick look, and then return the way we had come. Dylan reluctantly agreed to continue. He had been complaining he was tired and didn't want to walk any further, saying he wanted to head back to the car. I encouraged him to go just a little further. "Come on, let's just cross over and have a look, and then we'll head back."

Little did I know that last decision would nearly cost us our lives.

It only took a few minutes to reach the other side and walk the small hill that led to two paths. I told Dylan, "You choose which way. We will follow it to see where it goes; then we will head back." He chose to go right. We followed the path, noticing how it would slim down to the point the path nearly didn't exist, then go wider again. We walked around bends, crossed different parts of the dried-up creek, and then we'd be back to the path again. We followed it until it ended with an impenetrable wall of thick bush. So, there it was; we had officially completed it. With nowhere further to go, it was done. Yay! Now, I was ready to head back.

We began tracing our steps back the way we came, following the path, when we noticed the surroundings were unfamiliar. There were larger tree branches across the walking track we had to climb over that we hadn't clambered over before. We came across large open and flat areas, that we had not seen before either. I was confused. We turned around and walked back to the spot where the path had ended in the other direction, thinking we had overlooked the track we had taken, but no, it was the same. We went back and forth many times and kept returning to the same place.

My mind was racing by this time, and I began silently panicking, thinking, "WTF, what is happening?!" Mentally, I retraced our steps, methodically working backward while trying not to show Dylan I was starting to panic. "Remember we had to walk up a small hill after we walked across the dried creek river rocks?" I asked.

"Oh yeah," he replied calmly, believing that we'd just missed a little entry point from the hill, which would be easy to locate, if we were on the right path, that is!

Turning away from him so he couldn't see I was worried, I said, "OK, we got this," while my eyes darted back and forth frantically. I just could not for the life of me find the little path back to cross the dried river.

"Are we lost?" Dylan asked.

I felt a sickly feeling in the pit of my gut, but I said, "Noooo. I just need to walk a bit slower and look harder," trying to do my best to be confident. I could feel the air was cold, and looking at my phone, I knew we only had about 2 hours of daylight left.

I was really starting to panic and worry, thinking, "Shit, we are going to get stuck out here in the dark if I don't find the track!" Dylan was only wearing shorts, and I had my ¾ jeans on. Both of us only had T-shirts on with no jumper, no jackets. At that time, I hadn't even thought of having no food or water yet! Some water would have been handy as we had walked for some time.

As we continued walking back and forth, I faced the realisation that if I couldn't find the track, we would be stuck out here for the night. "It's only one night," I told myself, "We would be fine."

Still, I preferred not to, of course, more for Dylan's sake than my own. Where was that bloody path?! Desperately, I walked along, even looking across to the other side, willing it to pop out in front of us. The more I looked, the more it evaded me. "Come on! Where the hell are you?" I yelled silently, craning my head, trying to locate the wooden outdoor table seating.

I looked at my phone. "Shit, 10 percent battery life left!" I thought with more than a bit of dread. It would be bad enough to be out here for the night, but with no light or phone service, it would be even worse. My panic rose, and I could feel it begin to cloud my judgement.

In amongst my stressed inner chatter, I outwardly kept reassuring Dylan that we were fine and not lost. It was difficult as he could see on my face and hear in my voice that I was worried, which worried him too! At that moment, I decided to make a rash decision—to walk over the dried river rocks as close to where I thought the path was and climb the colossal mountain facing us. I thought it was the one that ran alongside the route that we followed down the track. If I could get up there and get service on my phone before the battery died, then at least we could call someone to let them know where we were so they could find us either that night or early the next morning. I knew we didn't have much time, so we stepped down onto the grey creek rocks, walked quickly across to the other side, stepped onto the grass, and started pretty much running up the mountain. The higher we got, the more brittle the trees became, and we needed to use them to help hold our weight to go from tree to tree to get higher. I led the way, step by step, instructing and showing Dylan so he could follow and know which tree to hold, when to sit, and when to crawl, while I continuously checked for phone service.

I believed that if we kept going higher, we would either come to a part of the walking track or the camping grounds or get service, surely. I was so fucking wrong. Completely freaking out in my head; my heart was racing so fast I could hear it thumping in my ears. The climb got steeper and steeper as we went higher up the mountain. I rechecked the phone—no service. Shit!

Puffed out, scared and unsure, I sat on the ground for a moment and told Dylan I would try to send Daniel a message on FB. Opening Facebook Messenger, I wrote my eldest son a message:

Daniel, this is not a joke. Dylan and I have lost our way back to the track we had been following...

My eyes began to fill with tears. Trying not to let the emotion get the better of me or alarm Dylan, I pulled myself together and continued to write:

We are in Mount Royal National Park. Call 000 to get us some help.

I clicked send. Nothing. I held the phone up higher and clicked send again. Nothing. I held the phone in every direction, clicking the send button. Nothing. Watching on, Dylan commented, "It won't send with no service, will it, Mum?" Being the voice of reason. I also had to face-reality as I tried to hold back the panic that had completely taken over my mind. "I don't think so," I admitted, but didn't want to believe it. I tried to click on our location on the map and use the directory to direct us back to the track, but each time I tried, a notification would appear saying there was no service available to detect the location. I was feeling beyond frustrated and said, "Fucking technology is so stupid."

From the clock on my phone, I knew we had, at most, 40 minutes before it got dark. I still believed we could find the track, but not from where we were. I decided that we needed to go back down the mountain to try and find the track with the remaining daylight we had left. "What if we find the track and need to walk back in the dark?" Dylan asked. I didn't even want to comprehend that thought, but I assured him that we would need to go slow and watch where we were walking as best as we could.

Making our way back down—it was steep. Trying to use the trees to hold onto was difficult as they were spread out, brittle, and breaking under our weight. It was nerve-wracking, and I was scared we would tumble down the mountain. To be safe, I thought the best option was to slide down on our bottoms. I would go first, with Dylan right behind me. That way, if I slipped down, he would know to stop, or if we both slipped, I would break his fall and try to use the trees as a stopper to slow our fall.

We continued down and were met with a whole heap of thorny bushes. We tried stepping through them, but their thorns scratched and cut into our legs and arms, stinging them so much because there were so many of them. It took at least a few minutes to get through the massive thorny patch of bush. I remember thinking how we never encountered them on our way up the mountain, which meant we were off course again! Shit! We needed to move to the right as we went down. Somehow, we had gone too far to the left.

Finally, we made it through the thick, thorny bushland, and with bloody cuts and scratches on our skin. A vast grassy ditch lay in wait for us, with a humongous fallen tree that covered most of the ground. I could see we would have to climb into the ditch and then climb along the tree trunk to get to the other, clearer side of the mountain, which I hoped would lead us down to the dry river rocks where we had started our climb up the mountain to get service—another bad idea in hindsight. One that cost us precious daylight time.

With about 10 minutes or so of daylight left, we wouldn't have time to climb along the tree trunk safely, so instead, we climbed into the ditch, puffed out and tired. As dark began creeping in all around us, I found some smaller trees and broke the softer branches off to use as blankets for us. Using my phone torch light with all of 4 percent battery life left, we made our bush bed for the night. I reassured Dylan we were okay as we laid down on the ground side by side. I tried to keep the mood light now I knew we had no choice but to stay put for the night.

"It's just like camping, but no tent or anything!" I said in an attempt to make him laugh.

In my mind, however, I was thinking, "Holy fuck! What have I done?!"

I positioned and patted the tree branches over our legs and up to our waist, trying to keep our bodies warm. We were both feeling the cold, but as we lay there, we talked a lot to distract our minds. We talked about how we may have missed the track, some of the twists and turns we'd encountered, and what we would be doing if we were at home. We mapped out a plan of what we would do the following morning, joked around, and did our best to smile, cuddling up.

It was cold and quickly becoming colder. It was also pitch black. We couldn't see a single thing, which was so scary. Any sound we heard had us absolutely shitting ourselves. I kept saying, "It's OK, nothing is coming near us," keeping him calm.

Meanwhile, my heart was pounding 100 miles an hour. I was terrified! We could hear so many animals and movement on the ground among the trees but couldn't see a thing. Trying to ignore the noises, I started talking through the plan. I believed we had only moved to the side of the mountain, so we needed to keep going to the right of it, and it would lead us back down.

Eventually, we both fell asleep. When I woke up to a few drops of rain on my face, I had no idea what the time was. "OMFG!" It's starting to rain! The dark and cold were bad enough, now, the rain began to come down. I felt so anxious. If we got wet while cold, we would get sick. How do I get shelter, even just for Dylan? Looking at my phone, I saw it had 2 percent battery left, and it was only 8 p.m. Dylan woke up and said he was cold and could feel the raindrops on his face. I needed to get up and start looking for more tree branches and grab whatever I could to make some sort of shelter for him.

Using my phone torch to see, I had only collected a few sticks before the phone died. Using the moonlight that had begun shining onto a few sections of bushland, I pulled some thick grass shrubs out of the ground. They were around the size of a dinner plate. I used the sticks to make some walls around

Dylan and the branches for the roof over his body, and then I laid the shrubs over the branches to make a bush-style tent for one in the shape of an igloo. I put my arms and legs close to Dylan's to keep him warm and sat in the doorway with a long stick as if I were his protector and guard.

Regardless, we both shivered all night, cold and uncomfortable. We continued to hear movement in the bush that surrounded us. I couldn't even imagine what was out there. My mind went on all kinds of wild tangents. Maybe I'd watched too many horror movies, but my terrifying fear of massive spiders and snakes made me feel like I was in my own horror movie—it was horrible. Dozing off, then being jolted awake to a noise, shitting ourselves, the cycle continued all through the seemingly endless night. I kept anxiously waiting to see the first signs of daylight throughout what I truly believed would be our one and only night. I was wrong.

Chapter 5

WE ARE LOST

Day 2 TUESDAY

Finally! The first signs of the morning emerged. I was already awake from being so uncomfortable, cold, wet, and terrified for most of the night. I was elated to see daylight come finally. I stood up, stretched my aching body out, and walked off to the side out of Dylan's sight as I needed to go and do that morning wee. He woke up shortly after. I guessed the time was somewhere around 5:45 a.m.

We waited until we had enough light to see our surroundings fully before trying to calculate which direction to move in. Allowing Dylan time to try to wake up properly, I didn't ask him too many questions. I already could see he was tired and cold. Instead, I kept reassuring him it was OK. We would get back on the correct track and go home to a nice warm shower, eat, and get into our soft, cozy beds.

I still believed that somehow, all we needed to do was go back up the mountain, then move further to the left to get back on the track and head back down the same way we came up. I felt pretty sure it would lead us back to the dried-up creek. It was beyond challenging to see anything other than miles and miles of thick bushland and a seemingly endless blanket of treetops. We couldn't even see the sky; it was so dense!

We started walking back up the mountain when Dylan was fully awake and ready. It was a vastly different start to the day than yesterday was, sitting in our warm, sunny kitchen, having fresh mango. The reminder made my tummy rumble. Water? I could only dream.

To describe what was around us is almost incomprehensible. There were many steep hills and ditches covered in thick grass, moss, and rocks littered with countless sticks, leaves, twigs, shrubs, bushes, small trees, tall trees, and tree branches that had either fallen to the ground or onto other tree branches. Some of the limbs were so big they would be nearly impossible to climb over, but we would need to, as we soon learned.

Climbing over the rough terrain, we held onto the stronger-looking branches for support while we climbed up or down the numerous slopes of the mountain. There were so many directions in which we could head, but we were focused on finding the path or track from the previous day so we could return to the car.

Heading up the mountain, we looked back to how far we had gotten from our morning starting point and realised it looked different from the previous day. My stomach dropped;

this way would not get us down to where I thought the trail to the car was. Standing still on this mountain, looking back in the direction we had come from, I explained to Dylan that I didn't think we would find the way we came in from yesterday. Everywhere we turned was uneven high and low levels of bushland—it looked very different to the bushland we had walked through yesterday.

We thought it was best to head back to our morning starting point and use it as a checkpoint, since it would potentially be closer to the track than any other. As we neared it, we both noticed to the left of us was another gully, and in the far distance. I pointed to what looked like a walking track. Dylan couldn't see from where he was, but I was adamant it was a track. Looking down from our position, we tried to plan how we would get there. It would be challenging, but with nothing other than thick bushland, it made better sense to aim for a clear track in the distance than keep persisting through the bush, not knowing where or when we would reach a trail.

The plan was to go down the mountain, walk over to the edge of the gully we had seen, navigate the drop we saw and the endless fallen trees and obstacles, and we should reach the track.

We walked down the mountain heading in the direction of the new gully, sometimes sliding down on our butts to reach the drop. Once at the drop, I jumped down first, then directed Dylan when he was ready to do the same. We had made it safely into the gully, yay!

Standing in it, we looked up at the edge we had just jumped down from and knew there was no turning back. It was a vertical drop and would be impossible for us to climb back up. Looking along either side of where we had come down, we were thanking our lucky stars we chose this part; because along the edge, the drop became bigger and bigger. A giant tree trunk had fallen right alongside the edge of the gully and drop. It had offered us a safer area for us to stabilise between the drop and the gully.

There was no such thing as walking a straight line through such thick bushland. We constantly zigzagged around massive fallen trees, branches, dense shrubs, and other obstacles while trying to focus on moving in the direction we intended. Eventually, we arrived at the location I had spotted, which up close was nothing more than a section of flat ground with different coloured leaves that made it look like it was part of a track. I was so disheartened. I began asking myself if I had wanted to see a path so badly that my eyes and mind had deceived and misled me.

Even though I was disheartened, I still had hope we would find the track from the day before. I kept thinking and saying, surely it can't be that hard! In my mind, we had only zigzagged a little too far to the left of the mountain, trying to return to the spot from the previous day. We couldn't have been that far away from it.

No thoughts whatsoever had entered my mind that this was only the beginning of the fight of our lives. It had been challenging until now, but I had no idea how much more dangerous, life-threatening, and unbearable it was yet to become.

There was no point in looking back. We couldn't climb back up from where we had come; in a sense, we had already tried that way. We continued through the bushland, trekking all day, climbing up and down hills, searching for a track or path of *any* kind. The only times we stopped were when Dylan asked to rest. What was once an adventure for him and me—began to lose its shine. He'd had enough, which I completely got, and he began to express more often how he was hungry and thirsty. It tore at my heart. I had nothing to offer my little boy other than reassurance. "It's OK, mate, we'll find the path soon and be out of here," I'd tell him, and I believed it.

I honestly wasn't sure how much longer we walked. The bush was like a never-ending, ground-hog kind of experience. I refused to acknowledge it or let it stop us from getting home. I can be determined and very stubborn. I don't like to give up or admit defeat. If there is another way, I will keep going. If I can't find another way, I will be resourceful and improvise. I've always had a solution or found one.

I talked to Dylan, pointed out different sights, shared stories, joked around, and anything else I could do in a bid to help him feel better about our shitty situation. Finally, instead of a sea of green, we spotted a dried creek bed. Although different to the one from the day before, it was a welcome relief from looking at bushland!

We walked nearer to it and realised as we stood at the edge of the bushland and looked down at the creek that we needed to slide down on our bottoms while holding onto the long grass and digging our fingertips into the dirt to prevent ourselves from sliding too quickly down the hill and into the dried creek. I remember thinking and saying to Dylan how there must have been a lot of rivers running through the park at some stage.

Once down the hill and into the creek bed, we stayed for a while. It was a welcome relief from all the climbing and bush, plus we really hoped we would find some water. Even better, we hoped it would lead us back to the original track and home sweet home.

Like the other creek, there were a million rocks of all sizes. This one had rocks from the size of a fingernail to the size of car tyres, and I thought back to my fears about popping a tyre the day before. I wished we had of now instead of getting lost! Some rocks were all bunched together, especially the smaller ones, but the bigger ones were more scattered and spread out. Some of the smaller grey ones were covered in dark green moss; some had moss only partially covering them. I thought they looked like black and white cows, except they were grey and green-spotted.

The creek was wide, so wide two cars could have parked side by side in it if it was flat, which, of course, it wasn't. Sometimes, we could walk freely, yet other times, we had to clamber and get down on our hands and knees to crawl when there were no gaps in the rocks to step through. In addition, just like the bushland, it was littered with broken branches everywhere.

Just like the national park seemed to be abandoned by visitors, this creek seemed to have been deserted too—of water, many, many years ago.

Feeling tired, hungry, and thirsty, we came to a dead-end cul-de-sac. I couldn't believe it; what creek ends abruptly? Saying we were feeling very disheartened and disappointed would be understating it.

Dylan looked at me, and I could tell he was waiting for me to tell him I knew what to do or had found a way out. He began to cry, and so did I. I sat and hugged him. My heart was literally breaking, feeling him sob in my arms, burrowed into me. He kept saying he wanted to go home, and I felt beyond devastated that I didn't know how to make that happen for him.

As tears streamed down his face and silently slid down mine, all I could do was tell him it was going to be okay. I began to look around, trying to focus on anything but the despair. To the left of us was a mammoth mountain, and in front of us was another huge mountain. Both had insane vertical and steep drops that even a professional climber would have likely found impossible to climb up, so those two options were discarded. Behind us was where we had come from, so the only option was off to our right. Even though it was super thick bush, it was our only choice. There was nowhere else to go.

By now, it felt like late afternoon was approaching again, and I knew in my heart we would not be going home today. I could feel the air was already starting to get cold, and I dreaded another night out here freezing, battling the non-stop shivering all through the night, not to mention having to fend off the fear of whatever noises we would hear. I decided to swap bottoms with Dylan. I was wearing ¾ jeans, and he only had a pair of cotton shorts on. With my longer shorts, he would at least be a little warmer on his legs and have less skin exposed to insects and other creepy crawlies during the night.

After the swap, I looked down and realised what a sight I was. Even Dylan, as sad as he was, gave a little smile. I looked like someone desperately trying to hold onto a youth that had long ago retired with my flabby, ghost-white legs squeezed into Dylan's shorts. I felt so uncomfortable and out of my comfort zone. It was the only time I was thankful no one could see me.

With at most a few hours of daylight left, I mentally switched into night mode and scanned the bulging green landscape for a safe place for us to rest for the night. I could see what looked like a huge white tree trunk so tall I couldn't see the top of it. It was in what seemed to be a tiny clearing that wasn't as thickly engulfed by bush. The daunting realisation we were spending another night spread across Dylan's face as I explained that we would get to the white tree and stay until the morning.

With heavy disappointment, he sadly answered, "OK."

I began learning very quickly the bush is incredibly deceiving. Nothing is what it seems. As we got closer to the white tree, we found it wasn't white, nor a big tree, just more bushland with a small opening to a section of flat land where we could see an opening to the sky. Everything else blended into the sea of green tree canopies towering over another shade of thick, green bush for as far as you could see in every direction. The only change was when they rose higher here and there on the massive mountains around us.

I was feeling worried, scared, and unsure. Dylan was frightened, hungry, and thirsty. It was a feeling worse than anything not being able to provide anything except my love, hugs, and words of comfort for him. Great uncertainty seeped throughout my body as I began to process everything properly. I didn't know what to think or feel, but I remember, in that moment, understanding and accepting that we truly were lost for the first time.

No matter what, I continuously reassured Dylan, providing hope that I would find a way out.

Walking and searching for a place to rest, I saw a ditch as big as a large car that had hollowed the Earth from the roots of a huge tree trunk that had fallen. The tree's base and roots were like something out of a movie. The roots were shooting out wildly in all directions. I remember thinking to myself, "OMG! What storm or winds would be strong enough to bring down a tree of this size? Fuck! How bad do the storms get in here? Did lightning strike it?" I didn't know and tried not to think about it because none of the answers were comforting. It was hard to fathom this huge ass tree just lying on the ground like it did.

Then I got an idea. I could use the branches attached to the trunk, the strong ones, and climb up on them to get higher up and see if I could see a way out. There was a clearing in the sky, and throughout the day, we had heard planes fly over us, but due to the thick bushland, there was no way we could have seen or been seen by them. Being in this location would give us a much better chance of being seen up high on the tree trunk. It wasn't much, but it gave me hope.

We grew even colder, hungrier, and thirstier throughout the late afternoon. I can't remember how often I apologised to Dylan—A LOT! I felt such deep regret and was riddled with guilt that I didn't decide to turn around when he said he'd had enough. Why did I let the stupid psychological babble get in the way? WHY!? Why was it so bloody important for me to finish a stupid track!? Who would even have cared if we did or didn't complete it? No one! Even more heartbreaking was when Dylan was crying and asking if we would EVER find our way out. I felt so responsible and questioned myself to great lengths about why I let this happen.

Dylan and I had different moments of crying. "Be brave and keep strong," I told him, being as encouraging as possible. "We WILL get out of here!" He didn't buy into my motivating talk. I didn't blame him at all. How could I ask him to trust or believe me when I had placed us in this situation?

Dylan didn't even answer with an "OK, Mum," as he usually would. Instead, he would reply with, "I'm hungry" or "I'm thirsty." The guilt and helplessness I felt ripped into me as I fought the tears back. I didn't want him to feel any worse than he already was.

The daylight was fading quickly. I needed to think clearly and be as rational as I could. We were here for another night. My hopes a plane would see us had dwindled. I had to get busy while I had the light and build some proper shelter to starve away the cold and protect us. I was fearful of it raining, even though the water would be great! The only water we'd seen since becoming lost was a tiny little puddle of murky, brown, disgusting-looking water in it. I got close enough to it that the foul smell wafted into my nostrils, almost making me hurl. To top it off, a few bees hovered close to it. I quickly discarded the idea of even trying to drink it, let alone allowing Dylan to.

I imagined it and knew it would be dangerous for us to get wet. We had been so cold the night before without even being wet that our teeth ached from the constant shivering for hours and hours on end.

I decided to use my phone cover to catch any rainwater I could. It was a cold and dry night, so at best, I could get some dew instead. I gathered as many branches from the smaller trees as I could reach and use. I remembered Bear Grylls making shelters out of nature; he'd made it look so easy. In reality, it was more difficult, especially with my long acrylic nails. The taller trees had bushier branches, but I couldn't get to them because they were so tall, so I broke as many branches off the smaller trees as possible to build a shack that would fit both of us in.

I told Dylan to sit on the ground to rest. I felt exhausted and drained of energy and knew his little body would be feeling it even more than my own. I didn't want him to use up any more of his energy or strength unless it was absolutely necessary. With Dylan looking on, I grabbed a few of the branches to use as a big broom to clear the ground. The night before, the ground was hard and uncomfortable, but it was worse with the sticks and small rocks that had dug into our backs, legs, and arms all night.

It had become dark by this time, so I used the bigger, bushier branches first, trying to stand them up and use the tips as support, leaning them onto each other to make a roof over our heads. It was working! I continued this for some time in the darkness, which had begun bringing the park to life with the sounds of the night animals. We could hear bushes rustle in the distance and sometimes close by.

With every sound, Dylan would ask in a hushed voice, "What's that?!" My heart pounded, but I calmly told him it was a kangaroo, possum, or whatever I could think of. The light from the moon went through sections of the bush and

the opening above us. It looked like a skylight in the clear, otherwise pitch-black night.

Using the light, I kept a mental map of the area firmly in my mind. The last thing we needed was to be separated! The shack that would be our home for the night was beside the fallen tree. I called it our outdoor seating, and then I pointed to another area and told Dylan that would be our bathroom and toilet area. Moving around in the locations where the moonlight didn't penetrate, I had to remember any ditches or large logs, avoiding them so we didn't trip over them during the night.

Dylan didn't leave the shack until daylight. He wasn't taking any chances. I continued building our little home for the night until after dark. Breaking off branches and filling as many gaps as I could with the smaller twigs, the walls and ceiling took shape under the watchful eye of Dylan. I think we both were amazed and shocked by what I had built, mainly because I'd never made anything before in my life. I thought it was the best I could do with what I could see within view of the shack. I had even built a little door!

Once inside, I sat beside Dylan. Knowing we had shelter to protect us for the night was a little more comforting. While we sat huddled together, we pulled our arms into our shirts to try and keep warm. As Dylan laid down, I removed my bra to provide him with some cushioning under his head. I put my legs over his legs to use any means possible to try and provide him with some extra warmth, but no matter what I did, we both were so cold. We shivered so much and for so long that the sound of our teeth chattering was louder than the sounds in the bush. Our bodies ached from the extreme and constant shivering. The night was the worst by far. It felt like it took forever for daylight to come.

At times, we both lay there on our backs, looking up at the night sky. I could see a red light flashing in the very far distance, but no sound made its way to us. I stared at it as it flashed and didn't appear to move. "What am I looking at?" I thought to myself. Then, I noticed dark shadows moving from high in the sky above the trees to below the trees.

I blinked a few times, trying to figure out what they were. I turned to Dylan and asked, "Can you see that?" pointing toward the red flashing light and the moving dark shadows.

To my surprise, he said, "Yes, what is it?"

Good, I wasn't seeing things.

"I don't know," I replied, continuing to look. I don't think I even blinked; I was looking that hard. We were looking at it so intensely that we forgot how cold we were for a moment, and our shivering stopped.

Transfixed, we saw the red flashing light, and the dark shadows continued to go from high above the trees to down low below them. I pointed and told Dylan to watch how the dark figure went from the top of the tree to lower, to out of sight below the tree. It was bizarre. If I had to describe it, it looked like army men were climbing out of a helicopter or something and shimmering down a rope to the ground.

The sudden realisation of what I thought it could be brought on an outburst of tears from me. I was afraid to hope, but I told Dylan, "I think it could be a rescue team!"

He began nodding and crying, too. The flood of relief, joy, and crying made it difficult to talk and get our words out.

"They must be near the carpark," I said.

"Will they find us?" Dylan asked amongst sobs. "How long will it take before they reach us? Do you really think we're near where we first started walking?" His mind raced, as did mine.

It had to be a rescue team; what else could it be? Surely, they would start looking for us on the track near the camping area where we had parked the car. I wondered how far it was from where we were. We had walked all that day, so there was a lot of ground to cover along the track, but I wanted Dylan to feel and keep some hope. I believed we both needed some positivity and to feel the light at the end of the tunnel.

We hugged each other, still crying and overjoyed. We laid back down and started talking about what we would do when they reached us, and the first thing we'd do when we got back home. Dylan said he'd go straight for the food. I said I'd have a shower. I explained to Dylan they would have water on them so we could have a drink as soon as they reached us. We watched more dark shadows slide down the rope for another 10-15 minutes, never taking our eyes off them. After a while, the red light slowly began moving in the opposite direction until it was no longer in sight.

Believing that a rescue team had been dropped off and had begun to search for us, the adrenaline of the excitement slowly faded, the cold seeping back into our bodies. We didn't care as much then, thinking it wouldn't be long until they found us. It was just enough hope to get us through the night. Dylan slept on and off. I dozed, keeping semi-alert to any noise, or periodically moving a little to try and find a comfortable position. There wasn't one. It's a big ask when you're freezing and lying on solid, cold Earth in shorts and a thin T-shirt.

I wrapped my arms and legs around Dylan, trying to give him any warmth I possibly had. For the first time in my life, I wished I had longer legs and arms and some extra weight on me to cushion my bones from the hard ground. My size ten physique just didn't cut it. Fuck, it was cold! I remember thinking, "What else can I do to generate some warmth for Dylan?" As his parent, I needed to provide and be resourceful, but this was beyond what I was capable of. It was terrible.

Throughout the night, I sat up and down, thinking about all the Bear Grylls videos we had watched and trying to recall his survival tips. One episode, in particular, came to me…Oh yeah! It was one where he was rubbing two sticks together to start a fire. YES! We weren't short of sticks, that's for sure. At first daylight, while we waited for the rescue team to reach us, I would make it my mission to find two small sticks, rub them together, and create a fire in the dirt.

My mind started to get carried away; not only was I visualising it, but for a split second, it was like I could feel the warmth of the fire. OMG, the warmth! I would be elated to finally provide warmth for Dylan. Then, the bubble burst. Wait, what if the fire got out of control? "Remember, we're in the bush, Michelle!" I scolded myself. It was bad enough that we were lost, hungry, cold, and tired, but worse would be running from an out-of-control fire. And the smoke! Oh…But then came another thought—the smoke! The rescuers would SEE the smoke and would know exactly where we were!

Throughout the night, Dylan would wake up and ask, "How long until they find us now, Mum?"

I put my arms around him and said, "Not sure. It's harder to see in the dark; we need to be patient and give them time to reach us, but they will." I said, mustering up all my confidence.

Chapter 6

STICKS AND STONES

Day 3 WEDNESDAY

At first light, I was woken by this continuous buzzing sound and an itchy feeling on my face, arms, and legs. I waved my hand around to shoo it away. The quiet lasted a second or two, and then the buzzing and tiny itches returned. Fucking flies! They drove me crazy.

I sat up, annoyed and frustrated, and tried to keep them from annoying and waking Dylan while wondering how we would get out. Fucking freeloading flies! Why were there so many? Didn't they sleep at night? I swatted at them, hoping like crazy they wouldn't disturb Dylan. The longer he slept, the better. At least while he was sleeping, he had some peace and wouldn't be thinking about his situation, how hungry or thirsty he was, and how much he was missing the comforts of his home, bed, and bedroom.

After a while, my body became so stiff and uncomfortable that I had to move, stand up, and stretch out. Slowly and quietly, I moved towards the opening of the shack, pushed open the branch-leafed door, and began climbing out. Once outside, I stood and stretched my body as people do when first waking up—but feeling far from rested, energised, and happy from a good night's sleep in a nice, comfortable, warm bed. There was no sugar-coating it, I felt fucked. This was fucked. This place was fucked. The situation was fucked. I hated it all.

It was odd, I thought, that I still started my morning by going to the bathroom for my morning wee every day. How I needed to, I didn't know. We hadn't had any fluids for almost two days now. Dylan didn't wee when he first woke up like he usually did at home; he was barely weeing once a day now.

"Mum, what are you doing?" I heard Dylan ask, slurring with sleep.

"It's okay. I'm just stretching my legs and going to the bathroom. You try to go back to sleep," I called, peering through the branches into the shack. He rolled over and went back to sleep, which I was thankful for. Sleep fixes everything, well, almost everything.

When the first rays of daylight came, my thoughts drifted back to my fire mission. I thought I'd try before Dylan woke up. I gathered small rocks to make a fire pit to contain the fire should I be successful, and a heap of sticks and dried leaves.

After numerous trips back and forth collecting, mainly the rocks, which in this area were no bigger than the palm of my hand, I had three different, healthy-looking piles. One for rocks, one for sticks, one for leaves. I sat on the ground in the dirt right next to the big hole from where the giant tree had been uprooted. The hole would have been the size of a large spa… "I wish!" I thought of soaking in a hot, clean tub of water. I'd settle for a bucket at this point.

I began positioning the rocks so they formed a complete circle, and then I picked up two small sticks and started rubbing them together hard and fast like Bear Grylls had done in the episode we had watched. I knew it might take a while. After all, he'd probably done it a thousand times, and no doubt made it look easy, but I believed I could do it, and the thought of Dylan waking to a warm fire was motivation enough.

After a while, Dylan woke up, stepped out of the shack, and sat down beside me. "Mum, what are you doing? Are you trying to start a fire?" he asked.

"Yes. Remember when we watched Bear Grylls rub two sticks together to start a fire? I thought I'd give it a go. I've got dried leaves and sticks for when the fire catches to keep it burning."

I was still going at it, even though my hands and arms were screaming sore. I felt like I had been rubbing the sticks for hours, and still, there wasn't even a hint of them smoking up. Dylan said, "I'm pretty sure the sticks need to be special ones."

I replied, "Well, I don't know anything about sticks except to make sure they are dead and brown, and that leaves are dead and brown too."

Internally, I was starting to think I didn't know what the hell I was doing. All I wanted to do was make a fire to keep my son warm, so I kept rubbing and rubbing and rubbing and *rubbing* for what seemed like another few hours—still nothing. "Fuck! This is not going to happen! How the hell did he do it? Did he have a hidden match or Flint or something? Maybe the camera crew slipped him a lighter," I said, which caused us both to start laughing.

My hands were all red, and my skin had begun to blister. I succumbed to the fact that I couldn't do it anymore and that no matter how hard and long I rubbed, it wasn't going to happen. I began to cry, and so did Dylan. I hugged him as we cried together. I kept saying over and over how I was so sorry. Him, being his amazing little self, said, "It's not your fault, Mum," which only made me feel like crying even more. What a beautiful, sweet soul this kid of mine was.

As the day emerged, we saw it would be beautiful without a cloud in the sky. It's all the better to spot a plane, hopefully. We didn't have to wait too long before we actually heard one. We both looked up simultaneously, stood up, ran to the tree trunk, and climbed the highest branch we could safely stand on.

Spotting a white plane, I began waving my arms like a crazy woman. It was probably ridiculous, but I believed someone was bound to see us since I was high up and flapping my arms around. Maybe the pilot or a passenger with binoculars on one of those sightseeing planes. I had to hope.

It wasn't a regular plane like the Virgin or Qantas fleet had. It was a small private plane commonly used for charter flights. Six planes similar to the first went over us throughout the day. From the ground, they looked the size of small toy planes. Each time we heard one, I would race to climb up the tree and wave my arms around, hoping and wishing someone, *anyone,* saw us. Dylan asked me, "What will happen when someone sees you, Mum?"

I told him they would let someone know they saw a person, notify them of our location, and send help.

Oh, how the movies portray things and give us false hope when, in reality, it couldn't be further from the truth. The gap from the trees to the sky was open and clear down where I was. Up high in a plane, however, the chances of someone seeing us through the peephole were slim to none. Even if they did see us, they would likely think we were waving at them, wouldn't they? I mentally went through the scenario. Who would know we were in distress or lost?

I thought maybe someone would have raised the alarm we were missing by now, and local sightseers or pilots would possibly be on the lookout for anything unusual from the sky. There would have to be heaps of sightseeing planes flying over Mount Royal. I learnt later that was not the case at all.

It was afternoon now, and as I had done the previous day, from watching the position of the sun and moon, I estimated it was roughly around 3 p.m. I encouraged Dylan to lay down and try and get some sleep while we had daylight, knowing that by the time night fell, it would be so fucking uncomfortable and cold we wouldn't get a decent rest. In addition to the elements, we were bloody scared. With every noise, fear rippled through our bodies like an electric shock. It was anything but restful during the night, and we needed sleep to think rationally and make safe decisions.

We laid down in our shack with the flies annoying the crap out of us and tried to doze off. Laying there, I don't know whether I dozed off or what, but my eyes snapped open at hearing a different sound. It was loud and distinct and getting closer!

I could tell it was far lower than the white planes because of how clear the engine sounded. I heard Dylan asking what it was as I rushed out with excitement to see which direction it had come from. "It's a helicopter!" I sang out to him as I ran to the tree. Scrambling up in record time, I couldn't see it, but knew it was close.

Dylan had raced after me but stayed on the ground next to the tree. Both of us wildly scanned the sky, feeling hopeful. He asked me, "Do you think it's the rescue helicopter looking for us?"

Nodding enthusiastically, I said, "Yes! It has to be!"

Closer and closer, louder and louder it became. We quickly scanned, our heads whipping around, trying to spot it.

Our hearts were beating out of our chests, and then we saw it! I felt like a child who had just gotten their only wish from Santa Claus after years of writing letters to the North Pole, finally seeing and holding that thing in my arms on Christmas Day! Nothing compared to it—the best gift in the world! That's how I felt, like all my wishes, but more importantly, the BIGGEST wish of all had come true!

The helicopter was not in full clear view, but I could see it, hovering off to the side of us. It was so close now. I could see that it was black, and I could even see the rotors. It was so loud and right there! In my line of sight! Within reach almost!

The hope and belief we could be rescued right then spurred me on to begin yelling, screaming, and waving my arms even more wildly than I had done spotting the planes. I even moved forward, back, and side to side, trying to create as much movement as possible to gain the pilot's attention. Desperately, I waved, willing the pilot to turn the nose of his helicopter just a metre more in our direction, and he would not only be overhead but also see us!

On the ground, Dylan was madly waving his arms, yelling, "We're here! Over here! We are here! Over here!" He waved his little arms with all his might, mirroring what I was doing.

I was frantically waving and screaming. I thought they HAD to see or hear us. "Here! PLEASE!" I begged loudly, thinking, *please* see us so that I can get my baby out of this nightmare!

I yelled until I was hoarse. My throat felt like razor blades had slid down it, and my arms felt like they were rods on fire. I was crying and yelling all at the same time, overcome with sheer desperation, standing on the fallen tree in the middle of bum fuck nowhere in Dylan's shorts with my Casper legs squeezing out of them.

My voice began to fade, and so too did my hopes as I saw the helicopter fly slowly and excruciatingly away from us. "No! Here! Please!" I sang out in one last croaky scream. Our hope faded, and in its place, despair and frustration took over. Still waving weakly, I felt like I was drowning with no lifeguard in sight. The tears didn't just well up; they bubbled up like hot springs and burst in a giant flood from my eyes.

Within seconds, the propeller noise got quieter as the helicopter flew further into the distance. I stopped waving my arms and attempting to yell out. Instead, I buckled over, exhausted, and held onto the tree branch, feeling absolutely gutted. I didn't move from the tree; I clung to it and the sliver of hope it would return, so I stayed put just in case it flew back. It never did.

After a while, I climbed down from the tree and hugged Dylan as we stood, hearing the last faint sound of the helicopter way off in the distance. We both couldn't believe how close it was. We were right here! It was right here! Why hadn't it just spun around, for fucks sake! It was strange. We felt heartbroken but hopeful at the same time because we knew someone was looking for us. They had to be! It couldn't have been just a casual flyover; they were looking for something, and I had to believe it was us.

We stayed put around the shack, cried some more, hugged some more, talked some more, and stayed alert in case it came back our way by some miracle, and I could waste no time getting up the tree again.

We fell silent.

A hundred times a day, my thoughts would drift to my other children, Daniel, Sarah, and Tim, as well as family members, my mother and father, brother, and sisters. I had one older sibling and three younger, all adults. I wondered what they were thinking—they would have to wonder where we were since we spoke so often and were a tight family.

I was concerned about my eldest son, Daniel, given he'd been through a bit with his last relationship, and I wanted to be home for him. I hoped he had a great time with his mate on the Central Coast during the long weekend, but I knew coming home to an empty house would have been confusing.

I hadn't told any of the kids what we were planning when I spoke to them in the days leading up to the weekend. I didn't know what Dylan and I would be doing until the day before the long weekend began. None of my other kids, family, friends, or workmates would have any idea where we were, or what we had been doing in the days before we became lost. I hadn't spoken to anyone other than Dylan since Thursday afternoon.

Now, being Wednesday, Daniel would have been home for a couple of days and would be worried about where we were. I felt he would also have alerted Sarah, Tim, and Mum or contacted my work to enquire by now. Being very close, I truly believed in my heart that for sure someone knew we were in some sort of trouble.

I would learn later just how bad it was getting for my other children and my dear mum by that day. Not only were they out of their minds with worry, frantically searching for us and trying to find out what happened, but they were also dealing with authorities, who, with no leads to go on, had suspicions of foul play committed by family members. The whole situation was a lot to deal with, and in some ways, I was glad I didn't know what they were going through; it would have been too much on top of what I was already trying to cope with.

Unaware of what was going on back home, I thought if Daniel hadn't alerted family, then my employer would be trying to contact me to see why I wasn't at work. I had never missed a day of work and turned up to every shift at least 10 minutes earlier. That's just the kind of work ethic I have, and the employee I am. I hoped they would think it was totally out of character for me not to show up.

As if reading my mind, Dylan broke my thoughts and asked if I thought Daniel would be out looking for us. "Absolutely, yes!" I answered, but honestly, I didn't know for sure. It had crossed my mind that maybe he had decided to stay longer with his mate on the Central Coast. If that was the case, he would have probably returned home yesterday during the day and wouldn't have thought it unusual that we weren't home since Dylan would be back at school, and I would be at work. It would only have been last night he would find it strange we weren't home.

We were always home and settled by 8 p.m. every night. One night would have been strange enough, but I knew two or more would have given him cause for concern enough to call his sister and brother to see if they had spoken to me. Once they couldn't reach me on the phone, knowing it was always on, it would have triggered more alarm bells for them all. I just kept telling Dylan they would all be out looking for us, but because we were in such a massive park, it would take some time to find us.

Throughout the rest of the day, I would scream out, "HELLOOOOOO!" hoping someone was walking on a nearby track within earshot. I asked Dylan, "What will we do if someone calls back?"

He just looked at me and didn't say anything.

"I reckon we would yell like hell! We are here, please help us!"

In my heart, I would feel relief that my child would be ok, receive food and water, and not be lost in this bushland in the freezing cold one night longer. In my head, I thought if someone did happen to hear me yelling out—please don't be a killer that's been hiding out in the bush to escape authorities. Yep, I've seen too many horror movies for sure.

The day continued in the same way—sitting, stretching, laying down, swooshing away the relentless flies, and waiting. When I heard the odd plane, I raced out, climbed the tree, and waved my arms at it, screaming. Other times I would yell out randomly, "HELLLOOOOOO!"

Surprisingly, the day passed quicker than it had previously. I kept busy, thinking that it was important for Dylan to see I was being productive and doing everything I could to get help and to get us out and that it would give him hope. I was giving my all and not thinking of how badly sleep-deprived I was, hungry or thirsty. All I could think about was that my child was lost here in this wilderness, and I must be strong for him, so all my mental focus was on Dylan and getting us both out of this god-for-saken place.

With the last of the daylight leaving us, I knew the hope of being seen and rescued before nightfall was gone. I felt bitterly disappointed as we climbed into the shack. We both sat in silence for the first time. I contemplated that we were in real trouble after so many nights of being lost. It dawned upon me that no one was coming to rescue us. None of the planes had seen us, and no one had spotted or heard us.

The sun went down, and the dread of spending another night here came. The darkness began enveloping us. Our hunger pains and feelings of extreme thirst had all but been forgotten with the distractions of trying to build a fire, spotting the planes, and the helicopter sighting. Now they returned all at once like a tsunami wave, with the realisation we were in for another night of fear and cold.

My heart broke a little more whenever Dylan said he was hungry or thirsty. Just thinking about it tore me up inside. I kept hugging him and saying, "I know, baby. Try not to think about it."

Yeah, right. It's an impossible ask of a child when food or drink is all they think of. My heart ached over and over, knowing I couldn't do something about it.

I didn't think much about food or water for myself. I only thought about it for Dylan. It was more so the water that worried the hell out of me. I knew how important water was, and if we didn't get some soon, we'd be in all sorts of trouble.

On the other hand, we would need food, but not as urgently as water. I had been over it a thousand times in my head and found it impossible to process how on Earth I would get food. We had no supplies like we would use if we went camping. There is nothing to kill or even catch an animal, let alone cook it! I couldn't even get two sticks to spark and light the fire we would need, even if we had supplies!

I went over every scenario I could think of. Could I make and set a trap? If so, how? And with what? Then what if we did catch something? How would we kill it? What if it bit me and it killed or poisoned me, and Dylan was left to fend for himself? It was a terrifying thought. I just hoped like hell we would get water soon, or better still, we were rescued, or we found our way out of this shit show before it became dire.

I tried to push the thought of all the creepy crawlies out of my mind. I had a complete and extreme fear of spiders in particular. At home, if I saw one of those eight-legged freaks, I cleared out. There's no way I would remain in a room with a huntsman, let alone any other kind of spider. The thought we were smack bang in the middle of probably millions of them wasn't something I had the time to entertain during the day, but it was a different story at night.

At every noise, I forced myself to think it was a harmless animal moving around like a kangaroo. We had seen droppings in the bush that looked like the kangaroo poo we had seen in zoos we had visited. So, with every rustle or noise, I imagined a kangaroo hopping or moving around with their joey in tow. I couldn't even think about a snake all slimy and hissing, slithering its way into the shack while we slept. I thought about it once and freaked myself out. I knew enough to know that should it happen, we were not to make any sudden movements—I had high doubts about that, though. I knew it would be more likely I would shit myself, scream, and jump around rather than lay quietly while it slid all over us. I'm pretty sure Dylan would have had the same reaction as me!

I was absolutely certain I would have screamed the whole Mount Royal National Park down if I had woken up to a tiny spider climbing on me, let alone a big one like a Tarantula. Tarantula! OMG, my skin crawls just thinking about or writing the word! I knew we were in the bush, and anything was possible, but in the dark, being blind to our surroundings at night it was a blessing and a curse at the same time. The thought of them was worse than the reality. All the scary movies we had watched in the past made our minds race with too many crazy thoughts in the dead of the night.

The shack realistically didn't protect us much from snakes, the cold, or my eight-legged nemesis. It didn't protect us from anything. I think all it did was give us a false sense of security. It wasn't much, but it was something, and at this point, anything was better than nothing.

It had become fully dark, and it wasn't easy to see anything as the moon was not as bright tonight. We could hear the bushes rustling, and honestly, it sounded like someone was walking through right near us, so I screamed out, "HELLOOOOO!"

Dylan said, "Mum, stop doing that."

I asked, "Stop doing what?"

"Screaming out," he said.

I didn't understand and felt unsure, so I asked him why.

He said, "It scares me every time you do that. I jump."

I put my arms around him and explained that I needed to so someone potentially passing by knew we were there.

When I thought about what he said, I could totally understand. It was dead quiet in the bush—no loud sounds, no music, no TV, no noise of people talking or walking outside, no traffic, no barking dogs, nothing, just silence unless we were speaking or making sounds. All the sounds we would normally hear in everyday life were gone, and it would make anyone jump out of their skin when dead silence is shattered with a high-volume scream! I told him I'd let him know before I screamed next so he could cover his ears and be prepared for it. From that moment on, it became known as the HELLOOOOO cue. He felt so much better with that plan in place. No more jump scares!

The night went the same way as the others had, shivering until our teeth and jaws hurt, our bodies aching from the cold and discomfort, and dozing until we were jolted awake by the sound of something close stepping on the dried leaves or poking around near us. Dozing, I woke in a panic, sitting bolt upright and freaking out, thinking I felt something nudging the branches of our shack door. Frantically, I yelled, "What's that?!" and began pushing big chunks of dirt with my shoes toward it to deter whatever was coming any closer to us.

Dylan was a whole lot calmer than me and said, "Mum, nothing is trying to get in." He's always been peaceful and the voice of reason. I took some deep breaths to calm myself back down and remembered that if I reacted like this, whatever confidence Dylan had in me would soon be gone entirely, which wouldn't help him or our survival.

"Breathe, Michelle," I told myself. Slowly and steadily, I reassured myself we were ok, and nothing was going to get us—well, I fucking hoped not! I assured Dylan I was okay and told him I needed to go to the toilet. In truth, I needed to step out of the shack for a moment because I was feeling overwhelmed with emotion from our situation. The immense guilt for getting my child into this situation, the constant concern I didn't have any food or water for him, no solutions or plans on how to get him out, and now here he was reassuring me! It was too much, and I didn't want him to know any of it. I could feel the tears coming and didn't want to worry or upset him further.

After a few moments of being outside the shack, Dylan sang out, "Mum?" checking to ensure I was still close.

I could only imagine how scared and lost he felt, and to feel alone in the dark as well. I knew I was afraid, and I was an adult! I couldn't imagine how much worse he would be feeling as a little boy. "I'm here," I answered, climbing back into our shack shelter. I could feel him relax, hearing and feeling me beside him again. He was instantly comforted by the knowledge I was still there and ok.

I sat back on the ground and cuddled up next to him, using my body to shield him from the cold. His little body shivered as he fell back to sleep. I felt so helpless. The brave face I had put on throughout most of the day kept the tears at bay, but now, as I lay there, they silently streamed down my face. I kept thinking, "Why didn't I listen to my gut feeling when we first drove into Mount Royal?" I had a bad feeling and ignored it. I had another bad feeling before crossing the creek. If only I had listened! It would have changed everything. The enormity of what I had done to my little boy was unbearable. I hoped he would be ok after all this. It deeply concerned me what effect this would have on him.

I cried for a very long time that night. Dylan woke up briefly, wrapped his arms around me, and said, "Mum, it's ok. We will be rescued."

I hugged my baby tighter as I felt him slip back to sleep. In my mind and the pit of my hungry, grumbling, empty stomach, I thought, "I really hope so," but I was losing faith.

I needed to rethink and come up with another plan. I decided no one was going to help us. We had to help ourselves and get out.

In every direction was thick bushland. With every step, we would have to battle through it. It was risky, whichever direction we took. I had no way of knowing what would lay in wait for us. Would it be a drop-off too big to handle, another dead-end creek, another dangerous mountain to climb, a wall of impenetrable bush, or a track at last? What the fuck to do?

Even in the full daylight, you couldn't see too far in any direction with the slopes and thick landscape surrounding us. During the day, I had whirled slowly around in a 360-degree fashion, looking and looking, and looking for a way out.

In my fatigued state during the day, it was hard to think. Directly in front of us was the big tree surrounded by bush so dense there was no way we could get through it. To our right was the direction we had come down from on day two, and that massive drop and cliff we couldn't climb back up. Behind us was bushland with tall trees that we may be able to walk through as it wasn't as thick, but I couldn't see too far, so had no idea what would be ahead. To our left was the steepest mountain of all. It was so high I couldn't see the top of it.

I knew our best chance was to get as high as possible so that someone in a plane or helicopter could see us if they flew over. We were partially hidden by the trees and bushland from down in the park's lowest part. We were so low we were perhaps too small for anyone to see us, even if they were looking from a plane. We had to go high but try and stay away from the cliff edge—no easy task.

Hurry up, daylight! I knew it would bring hope again, and I would focus all my energy on helping Dylan, preferably getting him out somehow. I *had* to get him out of here. For the remainder of the night, Dylan tossed and turned, woke up, and fell asleep. When he was sleeping, I felt better knowing he was relieved of the cold, fear, hopelessness, hunger, and thirst.

I huddled right up against his back, and with my arm around him, I dozed off, trying to keep him warm.

Drifting off to sleep, I imagined us climbing to the top of the mountain and reaching a huge clearing just as the first plane we saw came flying low above us. I saw us waving and the pilot waving back. I imagined several other planes doing the same in a short space of time, and ALL of them radioing in saying they had spotted us, and rescue teams finally arriving. It was a lovely dream.

Chapter 7

THE BIG MOUNTAIN CLIMB

Day 4 THURSDAY

Dylan's most restful sleep period was in the early mornings. Possibly due to the sun rising and the warmth embracing our face and skin. No longer shivering and feeling so cold, sleep was far easier to acquire. I looked at him and smiled, hoping his dreams were full of magical happy places and far away from this nightmare situation he was in.

Throughout the night, I had been thinking a lot about the climb we were about to embark on. I wondered how long it would take us to reach the top and how Dylan may struggle feeling so hungry, tired, and thirsty. I knew we would need to stop multiple times to catch our breath, and I would have to be super careful not to drain all of Dylan's energy.

His mouth had been parched from not getting any food or water, and his lips were visibly dried. I had tried hard to create spit and moisture in my mouth and showed him how to do the same, but even I struggled with it, so I could only imagine how hard it would be for a child of his age to do. The climb would be taxing, but if we had a hope in hell of anyone seeing us, we had to do it.

I remembered through the night an episode of Bear Grylls's show where he demonstrated how people could drink their own urine to survive. As disgusting as it was, and how much it completely grossed me out, we would be faced with having to do that if we were not rescued soon. I tried hard to remember the details from that part of the episode, and from memory, we could only safely do it once to add liquid and moisture back into our bodies to keep us alive. I wondered how we would do it, given we had nothing to capture it. None of the leaves or forage around us looked like it would do the job.

I looked at what we had with us; my car keys—forget it, my bra that had gone from Dylan's pillow to earmuffs, which were now cushioning his ear against the hard ground—might work. But the padding might soak up a fair bit, and then Dylan would have a smelly pillow—no, not an option.

Then I looked at my phone cover; it was one of those hard rubber ones with a hole in the back for the camera. I took it off my phone and at it, thinking, "OK, this is going to be the only way you're going to be able to catch any urine." Just the thought of it made my stomach lurch, but if it meant survival, it was a no-brainer.

Dylan was stirring and beginning to wake up, and I knew just being on the hard, twiggy, rocky ground was enough to wake anyone in a hurry from the sheer discomfort. I was glad he'd slept a little longer this morning. He would need the extra energy for what was to come.

As he became more awake, I sat beside him and put my arms around him as he sat up. I watched the harsh reality slowly spread through his face. Yes, we are still lost in this jungle, so to speak.

Feeling a bit anxious about telling him what we would be doing that day, I went through it in my mind first. Here goes…I asked him if he remembered Bear's episode where he was drinking his own wee. He paused for a few moments, sitting there not moving or saying anything, then looked at me and said, "What, are we going to do that?" looking visibly ill and disgusted. "Oh, yuck, Mum!"

I told him it was all psychological and we would need to put mind over matter. "We have had no water for days, and I don't know how long it will be until we get help. So, we are going to use the phone cover to catch as much wee as we can," I said, hypothetically demonstrating. "Using our fingers, we can cover the camera hole so what we catch doesn't leak out, then we have to drink it quickly. I want you to think of apple juice while drinking it, say apple juice, apple juice, apple juice, over and over again until it's all gone. Drink fast, no stopping," I instructed.

Looking at me, he was still shocked and surprised that we were even having this conversation. I said, "If it helps us survive, then we must do it."

He was looking at me like I was crazy. "Yuck! That's so gross!" he said, looking like he was going to vomit.

"I know, but just think apple juice, apple juice, apple juice, and drink it quickly. I will go first; how about that?"

Reluctantly, he said, "OK."

Already, my stomach was turning, and in my mind, I kept thinking, "OMG! OMG! This is the most disgusting thing ever! Get a grip, Michelle! Survival is survival!" This would be a last resort for anyone; you would only do it if it were a matter of life or death, and that's where we are at…Life or death.

I stepped out of the hut first, with Dylan following right behind me. With the phone cover in my hand, I thought, "Here goes," cringing. I said to Dylan as casually as I could so as not to freak him out any further, "Ok, turn around while I pee," trying to give us dignity at the very least. I pulled my shorts and underwear down and positioned the phone cover where I thought would be the best spot to catch the urine.

Using my index and middle finger to cover the camera hole, I began to pee into it. In theory, it was a whole lot easier than it actually was. Some wee went on the ground, some on my hand and fingers, some leaked out through the hole in the phone cover, but thankfully, some made it into it. I quickly and desperately brought the phone cover up to my mouth, closed my eyes, and repeated in my mind, apple juice, apple juice, apple juice, while my urine touched my lips and went into my mouth. It was warm, and as I swallowed, I began to heave. I shut my eyes tighter so I couldn't see or think about what I was drinking, and the taste was fucking worse than ever. It was foul!

I couldn't even give you an example of how disgusting it was if I tried. There was nothing worse or that I could even compare it with. I pulled my underwear and shorts back up, shaking off my hand and the phone case of any excess urine. Using my shirt, I wiped it clean and handed it to Dylan while hiding how bad the experience was from my face.

I said, "Ok. Now it's your turn," smiling to give him a sense it wasn't all that bad and he could easily do it too. He was reluctant to even take the phone cover. Knowing I had just used it to capture my urine, he wasn't keen to even touch it! I didn't blame him.

I held his hand to the phone cover and put his fingers in place to cover the camera hole. In his case, he needed to use three fingers to plug the hole. I said, "OK. I will turn around while you wee into the cover, then remember, as fast as you can drink it, and keep thinking apple juice, apple juice." His face was pure disgust as I turned around.

A few seconds later, I heard urine hitting the ground. I quickly turned around to see him struggling to block the camera hole. He was losing urine fast, so I said, "Quick, drink what you can! Apple juice, apple juice, apple juice," I chanted as he put it to his mouth, took one taste, spat it out, and threw the cover.

"No way! That is so yuck! I can't, Mum," he said between spitting, trying to remove the taste from his mouth.

I said, "It's ok. I know it tastes nothing like apple juice. If only."

He was still spitting minutes later. I completely understood. I will tell you this; it might be a taste I may not be able to explain, but I will never forget it. The aftertaste lingered for a long time afterwards and made me feel like I wanted to vomit.

I had the car keys already sitting on a tree so I could put them back into my jean shorts that Dylan was still wearing. I didn't want to risk losing them or my phone, so I carried both in my hand. I honestly thought that when we managed to find our way back to the walking track we initially started on, we would be able to follow it back up to the camping site where I had parked the car and jump in it, plug in the phone to charge it, call my other children and let them know we were ok, then drive out and go home.

The sun was starting to beam brightly and was rising higher in the sky; we needed to start getting up the mountain! At a guess, it was around 8 or 9 a.m. I remember learning once you could tell time in nature by the sun's position. Noon was when the sun was directly above at the highest point, and it would set at the 6 p.m. position. It was 9 a.m. when the sun was halfway up to the 12 o'clock position and 3 p.m. when it was halfway down, heading down to the 6 o'clock position. It wasn't quite at the 9 o'clock position, so it was after 8 a.m. and close to 9 a.m. Being able to estimate was handy to ensure I had enough time to get ready to settle for the night.

Looking at the huge size of this mountain was daunting. It loomed out of the Earth like it was a planet itself. I took a deep breath, "OK, let's go," I said. To begin with, it was steep but not so steep we would slide straight down. We often couldn't walk normally; we had to bend over, grabbing onto shrubs and trees with our asses pointed at the sky the further we climbed.

Dylan stopped to sit and rest often to gather the energy to move on while I remained standing, thinking I may not have the energy or desire to get back up again if I sat down. Below and some distance away now, I could see the shack and suddenly remembered I'd left my bra and the key ring that had fallen off my keys from Dylan's first day of preschool there. Dylan looked at me and asked with dread, "Do we need to go back?"

No fucking way we were going back down. "No, it's ok. We will keep going, but you won't have the bra to help keep your ears warm and use for a pillow anymore."

He shrugged his shoulders and said, "Oh well." He was so exhausted that he didn't really care, and the relief on his face confirmed he was totally ok living without his pillow.

We climbed, we stopped, climbed, and stopped for hours. We could no longer see the shack from our vantage point to the left. Instead, we saw the tops of the bigger trees in the park. We were making great progress. Dylan asked to stop and rest. We caught our breath, sucking in as much oxygen into our lungs as we could through our dry mouths. Trying to keep them moist was becoming more and more impossible.

I felt buggered and decided I'd sit on this stop as well. I plonked myself down on the hard ground and felt a gush of wetness. "WTF!" I thought, wondering if I had just wet myself. Confused because I hadn't even felt like wanting to go to the toilet, I looked down at my shorts and couldn't believe my eyes. I had got my period.

"You're fucking kidding me! What else?! Can this get any bloody worse!" I yelled, ranting and raving and throwing my arms in the air.

Dylan looked at me, unsure of what was happening and why I had suddenly begun yelling. "You ok, Mum?" he asked, worried and surprised at my sudden outburst. I think he thought I had lost my marbles!

I dropped my head between my knees and began to cry. I couldn't take much more;-our situation was bad enough. Every minute, every moment was challenging, trying to find and explore the best way to get help and save my child from having to spend another second in these conditions. Now, to make it even worse, I had to deal with my period with no supplies, change of clothes, access to a shower, and also figure out how to explain it to my 9-year-old child who had no idea what a period was!

I thought about animals smelling the blood scent and trail and how that could put us at risk of attack. I kept my head down, thinking and crying. Dylan placed his arm around me. It offered so much comfort in that moment and was the perfect action to help me snap out of it and remember to think of him. I raised my head, wiped away my tears, took a deep breath, and launched into the facts of life about being a girl.

When I was finished, I scanned his face. You could see he was more than a little confused and shocked to know females bleed from their vaginas. I felt he was too young for this conversation, but there wasn't a choice. He asked, "Do boys do it as well?"

"No, but I wish they would, so they understand the experience and what we go through. It's not pleasant. Even with all the supplies, pain relief, fresh clothes, and showers, it's still uncomfortable, sometimes messy, stressful, and tiring for a female."

I realised I would need to change back into my jeans as they were thicker and would absorb the blood better than Dylan's thin cotton shorts. They were also tighter around the legs, so would catch any blood leaking down my legs better. I felt terrible having to swap with him. He really needed the jeans to keep warmer on the ground at night. Explaining why I would need to wear them again, Dylan didn't hesitate for even one second.

We turned our backs to each other, peeling off our shorts. It didn't worry me if he saw me in my knickers, but I knew he was coming into that age group where you would rather die than see your mum in her underpants! We handed each other our shorts back and redressed. My legs appreciated the warm feel of having more of my legs covered, even though I was so upset that I had to change with him and take away that little extra comfort and warmth they provided for him.

I felt like I was ripping him off and felt more deflated and defeated than ever. I needed to think of something to help with the blood loss. I had to improvise and be resourceful as best as I could. Leaves, maybe? I doubted it; the blood would run straight off of them and wouldn't absorb. While sitting there, the thought of drinking my urine that morning popped into my head, and I thought, thank goodness I had done it this morning! As disgusting as it was, it was better then and not now.

Then it occurred to me that I could use parts of my T-shirt! It was the only item I had that could at least do a bit of the job. Where should I start? If I took it from the bottom of the shirt, it would make it shorter and harder to pull over my knees at night when trying to keep warm. I needed to be strategic and try to make the best possible decision. Maybe the sleeves? I didn't have to spend time thinking about that. I needed to pull my arms in from the sleeve at night to keep warm too…For fucks' sake. I had to choose one, and so the sleeves it was.

I tried to tear the sleeve, pulling and pulling. It's not as easy as they make it look on TV. No matter how hard I pulled, I could stretch it but not rip the material. Fuck! Now what?!

Dylan sat beside me, watching what I was doing while I awkwardly tried to explain why I was doing it. I couldn't figure out if he was becoming more confused or didn't care due to our circumstances.

Sitting on the ground, we could feel sticks and rocks underneath our asses. Flicking the larger rocks and sticks to sit was becoming all too normal, and we had done it without even thinking about it. Being on the ground all night and off and on through the days, we had started to ignore how uncomfortable it was. I didn't even care about the lack of sleep. Most of my mental processing went into figuring out how to get out and how to protect Dylan. "Mind over matter!" I said, often to get me through.

Throughout my career in Hospitality, I could work so many hours, and, on some occasions, 7-day weeks. In general, I had never been a good sleeper. I had difficulty falling asleep and staying asleep and was notorious for continuously waking up throughout the night.

Being an early riser, regardless of the lack of sleep, it was nothing for me to soldier on through the day, completing tasks, getting through the workday, and spending time with my kids after work and on my days off. I felt like I functioned like someone who slept 7-10 hours a night. I used to think, wow, if I can accomplish what I do on minimal sleep, imagine what I could achieve on the recommended sleep! I never realised just how exhausted my body and mind was.

For most of my life, I was on autopilot, defaulting to just doing the do and getting what I had to do done. There was little time to stop and smell the roses, little time for being present and appreciating even small things like food, water, my morning cup of go-go juice, toilet paper, toothbrush, clothing, bed, amenities, and toiletries like pads or tampons.

On this day, I thought about all those things and more as I focused on the task of putting some absorbent cloth in my underwear. I could feel period blood trickling down from inside of me and making its way out.

Again, I forgot about and dismissed any body pains or how tired I was in lieu of wracking my brain to devise a way to cut my shirt. Then I remembered I still had the car keys in my pocket. I dug them out and used them to put a hole in the right-side sleeve I had been pulling at. Pushing into the cotton material hard, the key formed a small hole. I pushed my middle finger through to make the hole bigger, then grabbed both sides of the hole and pulled hard, ripping the sleeve free of the T-shirt. The sleeve wasn't large, but it would be enough for its purpose, I hoped.

I stood up, turned around, and wedged the sleeve between my vagina and underwear. It felt so bulky and weird, but then again so do pads—uncomfortable at the best of times. The cloth would do; we needed to get going. "Let's go," I said to Dylan.

He was reluctant and slow to get to his feet. Crouching, we resumed working our way up the mountain. The higher we got, the steeper the terrain became. It taunted us with each move we made. Our hands were filthy. Our fingers and fingernails were covered in dirt like young children playing outside in the dirt making mud pies. Strangely, I still had all 10 of my acrylic nails intact. I was already overdue to have them infilled before entering Mount Royal, so the growth between my natural nails and the acrylics was now sealed with a thick layer of compressed dirt.

The bush tried to deceive us constantly. Looking back briefly in the direction of the shack, it looked completely different. If I had been any less meticulous about our surroundings left, right, below, and above, I would have thought we had taken another path. This bushland seemed to evolve and change behind you, so you couldn't return in the same direction you had come.

We had no way of knowing how far up we had come, but from the sun's location, I knew it was around midday, so we had been climbing now for around 3 hours. We were both tired and extremely thirsty. I could hear both of our bellies growling loudly for food. Dylan didn't even bother asking for food or saying he was thirsty anymore. It took all my remaining energy to hold back the tears from not being able to provide for him. I stopped to hug him and offered all I had, which were words of encouragement to keep him going.

After a while, as we got higher, we came to a section of the mountain that was sparser of large trees. There was an opening to the sky, but no sign of the top of the mountain. Again, there was more trickery from this never-ending green and decaying mass of different-sized trees and dead and alive grass, vines, and shrubs. The terrain was a maze of it all, with patches of dirt, fallen trees, branches, and rocks of all different sizes. It was like we had been transported to a different world. If a dinosaur walked out, I doubt we would have been surprised. It was huge in scale and peppered with massive mountains and bush sprawling out in every direction. It exhausted me just looking at it.

Spent, I decided this was where we would stop and rest for the night. We chose our toilet section, even though there were only number 1s now, and they were few and far between. Having my period, I assumed I would need to toilet more often, but without food, I wasn't sure. I planned for it anyway.

In front of us was all uphill; to the left was a large area of shrubs; behind us was the steep decline; and to the right, there was a thin dirt path leading to a small flat patch. We joked around, calling the path our hallway. Around the shrubs was another dirt track the size of a BMX wheel; it led to a huge tree as tall as a 5-story building. Its circumference was so wide, that eight people could stand with their arms outstretched, touching fingertips, and they would barely surround it.

Entertaining ourselves, we used our imagination to section off different areas, nominating where our imaginary living room, TV, and bathroom would be. Above the flat dirt patch were branches from two other trees interlocked overhead that provided a roof of sorts, and the large shrub area we decided would be our garden. Playing this little game not only appealed to my 9-year-old son but also helped us both have a sense of connection with normal, everyday life.

We built our bedroom while shooing away the annoying flies that had increased in numbers, attracted to my period blood. This was yet another gross piece of information I had to convey to Dylan when he asked why there were so many.

I could tell Dylan needed sleep, so I encouraged him to lay in the dirt patch in our 'lounge room' while I kept creating our shelter for the night around him. He was so buggered, and he had no energy left to protest even if he wanted to. As I walked around the trees and shrubs looking for materials for a shack, I would hear him ask, "Mum?" to ensure I was still close when he lost sight of me.

"I'm here," I'd reassure him, moving back into his line of sight.

Sadly, it was now normal for him to doze instead of falling into a deep sleep. Fearful of being separated, he would doze and frequently open his eyes to make sure I was there if he couldn't feel me sitting or lying beside him. I couldn't even imagine how terrifying it would be if I lost him, let alone how he felt about losing me out here.

Searching the shrubs for some foliage, I noticed they were very similar to the ones I had used on our first night to build the igloo shelter. The roots of the shrubs were the size of a cereal bowl, and the grassy weed-like foliage was the size of a dinner plate, making most of them relatively easy to pull out of the ground to make a roof. I shook off the excess dirt so none would fall into our eyes and stacked them up next to where Dylan was dozing. The last thing we needed was dirt in our eyes. I always had to think of the worst-case scenarios to avoid more challenges.

I felt like I was a farmer or gardener; mind you, in my regular life, I was neither. I didn't do any gardening or farming at all. The closest I would get to either was mowing the lawn, and I don't think that counts!

Our surroundings were so quiet there was barely a sound. I took a good look out into the distance, and it hit me hard how fucking massive Mount Royal was. In every direction, it loomed. There was nothing else but Mount bloody Royal, with its thick bushland and endless mountains.

Seeing it from this vantage point sent me into a bit of a panic. It was much bigger than I could have ever imagined. How the hell was I ever going to find a way out? All these mountains! I didn't even know if we were on the same one we had started on. My mind began racing as discouragement, and disbelief overwhelmed me. We were literally a dot, a microscopic one in this place. The chances of being seen, let alone rescued, were slim.

I estimated that Dylan dozed, woke up to check I was still close, and dozed back off constantly for an hour or two. Having collected as many shrubs as possible, I began placing them one after the other on the branches to cover them with the weed-like shrubs. I thought about making a bed of shrubs or leaves but thought it would only attract creatures during the night and didn't want anything crawling on us. It may not have happened, but I took some comfort in believing it was because of being so fearful of snakes and spiders.

After a while, I could balance and stabilise the shrubs so they weren't falling off. There wasn't much support from the branches because I had to use the thin ones. We didn't want heavy branches falling on us during the night if the wind came up.

Once completed, I felt a slight sense of accomplishment in building something to protect my baby, even if it wouldn't protect us from a storm or heavy rain. In my mind, we had a barrier between us and whatever lurked in the bush. What I lacked in bush skills, I made up for in improvisation and determination, with Dylan being a powerful driver.

It also made us invisible to a degree from someone who was up to no good. The chances of a stranger coming randomly across us were slim unless it was a rescue person. I thought the chances of someone wanting to harm us were even more unlikely, but it still played on my mind.

Watching Dylan as I moved around our area collecting more shack materials, I saw signs of movement as he began waking up. I had been watching him like a hawk—ready to kill by any means necessary if anything was to approach him or attempt to hurt him. He sat up, looking around in search of where I was, and I could see him stiffen a little with that first realisation of where he was, followed quickly by the fear of me not being close.

I quickly sang out to him, moving so he could see me. He shifted slightly, trying to work the uncomfortableness from laying on the hard Earth out of his body. Night after night, the toll of laying on small rocks, sticks, and the ground was getting worse.

It amazed me how our surroundings were the same, but the terrain could change so much from place to place. Thick, impenetrable bush on the flatter ground one minute, to sparse bush and steep hills the next. Massive boulders in one place, and a billion small rocks in another.

The one constant was the dried creek beds, some with rocks covered in moss, but never a hint of water. It was beyond frustrating. A part of me was so angry, purely because it was the one thing Dylan needed for the best chance of survival. I asked him how many years ago he thought the rivers were full and running through this land. We sat imagining them abundant with fish and clean, beautiful water. Thinking more on it, in a way, it was good because if they were full, we wouldn't have been able to get across them at several of the locations we had crossed, but we would have at least been able to drink the water.

I sat thinking that if we had to swim across the creeks, we would have been wet and then would have had to battle the freezing cold nights in damp clothing. It was hard thinking about it, let alone doing it. The pain we experienced dry at night, and shivering was already high. On a scale of 1 to 10, enduring the night wet would have been an unbearable 10+ at best, and we would have ended up sick—one of my worst fears was one of us getting sick or injured.

The sound of a small plane broke our thoughts and conversation. We had seen and heard a few. By now, I could tell just by their sound if it was one of the small white planes or a helicopter. Being in a larger, more open space and being higher up, we hoped someone would see us waving our arms about and jumping up and down. We couldn't give up hope.

Being busy creating the shack, I had all but forgotten about my period until I felt another gush of wetness fill the bush-made pad as I jumped around, waving at the plane. The lack of personal hygiene was a real worry, and I knew I'd have to make another pad. Using my car keys again, I started tearing at the left side sleeve in the same way I had the right side previously.

Freeing the sleeve from my T-shirt, I went to our bush bathroom area behind the tree and changed the cotton-made pad. I was still worried the blood might attract some animal and wanted them to stay far away from us, so I walked a small distance away and threw the blooded wet cloth into the bush in the opposite direction of where the shack was. I hoped it would be far enough away. I couldn't go further than that because I would lose sight of Dylan, so it would have to do.

I was feeling exhausted and needed to sit and rest. Being beside Dylan gave us both a sense of comfort—we were all we had. Throughout the afternoon, I would sit and listen to Dylan talk and ask questions.

We daydreamed about how many people would be searching for us and when and where we would be rescued. I thought it would be a team of 4 to 6 people, and Dylan thought more, maybe 10 to 20 people with search dogs, too. We surmised whether they would be on foot, by plane, or by helicopter. We talked about how we would jump for joy and feel so happy no matter how they came, as long as they did! We thought we would be walking through the bush, hearing them calling out our names, and having backpacks with food and water for themselves but extras for us, too.

We would talk about seeing four helicopters circling us in the sky and winching down a rescuer to retrieve us. We even had a contingency plan if the bushland was too thick to winch us up! We thought they could do a food, water, and blanket drop and fly low and slow, leading us to a safe area where they could pick us up from. Dylan's face lit up, coming up with a host of rescue scenarios. I honestly didn't care which one it was, as long as one of them would eventuate, and soon.

The talk distracted Dylan from his reality and the feelings of hunger and thirst. I never spoke of food or drinks unless he brought it up, mostly because I didn't want to remind him or trigger his hunger or thirst, although I didn't stop him when he spoke about the foods he loved or what he would eat first when we got out of the Mount Royal National Park. We did a fair bit of drooling, talking about foods we loved, like steak, prawns, Chinese food, and the mango smoothies we made and loved to drink. OMG, yum! How nice would all those items be right now, we dreamed. I closed my eyes and laid back on the ground, thinking of them all. I could almost taste them in my watering mouth as I shooed away the annoying fucking flies.

I had no idea how rescues were carried out. The most I knew was from movies, and even that could have been embellished. After all, I thought tearing a sleeve off a T-shirt would be a walk in the park, and it wasn't. Only movies and TV shows make things look easier than they really are.

We lay in silence on and off. Perhaps because we both knew no matter how much we spoke about food and drinks, it wasn't going to make it appear and fill our hungry, grumbling bellies.

As darkness descended, I held on tight to the phone, forever hopeful that if we found our way back to the car, I could charge it. I felt for the car remote in my pocket and clicked it, hoping to see the flash of the purple LED lights that were along the side steps of my car. Every night, I did the same, pointing the remote in all directions.

I had no idea which direction the car was in or whether it was near or far. We had covered so much ground, climbed many hills and mountains, and trekked through so much bushland I doubted we were even in the same area as the car, but you had to hope.

The cool air wrapped around us, and the dread of another freezing, frightening, and uncomfortable night seeped into our minds. Soon, it was pitch black, and the sounds of night came with the bushes rustling around us. It was scary, but the prolonged freeze for hours and hours was worse.

Laying on the dirt, I thought to myself, "Here we go again—another night freezing our asses off and shivering until we hurt." It was worse tonight because I felt the immense guilt of having to take back my ¾ jeans from Dylan to stop the blood running down my legs.

Dylan tossed and turned, trying to sleep. It wasn't about being uncomfortable so much; it was because of how cold he was. The constant shivering was relentless. Our bodies and jaws ached from shaking. His thin cotton shorts were no match for the nights here. I huddled up closer to him, trying to give him as much warmth as I had in my body.

Unable to sleep, the night dragged on. The only relief was watching the moon orbit the Earth through the trees. From its movement and position, I could tell how close to morning it was, but it was like watching water boil; the longer you watched it, the longer it took.

Strangely, I could only hear bush movement from a distance throughout the night. Maybe it had something to do with us being higher up where the animals didn't climb? I didn't know, but it comforted me, knowing they weren't as close as they had been. Listening, mostly all I could hear was crickets and a dog…What?! A dog?! Was I going crazy? I laid still, listening hard. Yes! It was a dog barking. It was faint and sounded far away, but I could hear it. I turned to Dylan, who I knew was awake too, and asked, "Can you hear that?"

We both sat up, rolled out from beneath the shack and stood up. It was the first sound or sign of civilisation in 4 days. I walked slowly, feeling my way in the dark toward the direction of the barking. We were both quiet as mouses listening. Then we heard the dog again. This time, Dylan heard it too and said, "Yeah, Mum, it's a dog!" looking into the distance. We couldn't see anything but darkness.

We peered intently in the direction of the barking, and I could see a very tiny orange light. I wondered if it was a porch or outside light; it looked like it to me. Even though it was a long way from us, it offered us hope there may be a house or houses in the mountains ahead. I thought immediately of how we could get there. We were high up, and the house seemed far away but not too low. "Surely there would be a track through the bush near it," I thought aloud. Dylan, who sometimes offered his input, said nothing, preferring to contemplate silently. One thing I knew was it wouldn't be a nice, straight, direct line. The chances of us going off track to move around obstacles would be high, as we had already learnt and encountered.

I stood still, with my arm around my boy. We looked on, fascinated by the light and the faint dog bark. It was comforting to listen to a familiar sound from home. While we couldn't trek towards the house tonight in the dark, I knew we were closer to home than we had been at that moment, and it was a relief.

I told Dylan there was nothing we could do during the night but come morning; we would see if we could spot anything in the direction of the light and dog. We laid back down, and my mind was busy for the remainder of the night. What were we looking at? Was it a house or a cabin? Was it part of the Mount Royal or a house along the road that led into the national park? Would we be able to hear the dog tomorrow to lead us there?

Dylan slept on and off and would occasionally ask his own questions or say, "I can still hear the dog barking, Mum."

To pass the time, we'd try to guess the dog's name or why he was barking so much. We wondered if it could be part of the search team looking for us. Maybe the light was for us to see? Who knew?

Chapter 8

REST DAY

Day 5 FRIDAY

The flies at first daylight, swarmed us like bees to honey. We were sleep-deprived and not in the mood for them. I swooped my arms through the thick of them, saying, "Go the fuck away!" Using the shrub bushes, I tried to cover our faces and skin to provide some relief from them, but it was useless. They were persistent little bastards. My mouth was as dry as the creek banks here, and I felt exhausted. I licked my lips to try and moisten them while shooing the flies.

Unfortunately, now that the sun had come up, the dog had long since stopped barking, and the faint orange light I had seen in the bush the night before was gone entirely, and nothing could be seen in its direction. Without it, it would be impossible to locate exactly where it had been, and with the bush and terrain between us and it, walking blindly in its approximate direction wasn't going to work either. It was far too easy to get off track. We could end up on a wild goose chase and worse off than we already were.

Ultimately, I decided with the openness to the sky in this little section of the bush, it would be best to remain here for now.

Dylan was still dozing in and out, his lips dry and cracked. He had begun complaining a lot about being thirsty and hungry and was crying a lot more, asking if we were ever going to get out. My heart broke, and I hugged him with tears in my eyes. "I hope so," I replied, silently wishing one of the helicopters or small planes had seen us and a rescue team was close. Just in case, I would not let any chance pass me by when I heard a plane coming towards us.

Sometimes, Dylan would beat me to it and yell, "Mum, another plane!" and jump up and down, waving his arms frantically and desperately wishing for someone to see us.

Each time I stood up, I could feel I was becoming more tired. I was extremely low on energy, to the point my body was saying no more, but my head was saying, "DO NOT GIVE UP! Get your child out of here!" It was all the motivation I needed to keep pushing through it and focus on connecting my brain to my body to get myself off the ground and coordinate my arms to wave in the air.

Being much smaller than me, Dylan would be burning up even more energy. Until now, he'd never worried about having food or water; it was just there, and I always ensured our fridge and pantry were full. For him to go without a meal, let alone five days without anything, was extremely difficult to comprehend. It was a lot on top of the situation we were in. His willpower amazed me, along with how resilient and remarkable he was, holding himself together, and me too at times!

When he raised the topic of food, particularly his favourite meals, I cringed, thinking it would make him even hungrier. However, working and keeping his mind occupied, strategising and thinking ahead was good for him.

Late afternoon approached, and the sun no longer warmed our skin. The familiar disappointment of not being rescued set in. -Using fern leaves to try to cover our faces to get relief from the flies and some rest, we chatted about all sorts of things to take our minds off the constant cramping that had begun in our stomachs. Our internal rumbling was louder than a thunderstorm.

Both Dylan and I got up in the last light of the day to use our bush bathroom and returned together to our bush bedroom. The blood from my period had now seeped into my jeans, and it felt horrible and uncomfortable.

We lay on the ground in the dirt, once again huddled to keep warm. It makes you really appreciate your floors, walls, bed, toothbrush, toilet paper, pillows, blankets, pads, shower, electricity, food and water, everything we use in our everyday lives, usually without any thought, was gone.

Chapter 9

THE WINDSTORM

Day 6 SATURDAY

The first glimpses of morning light came, thankfully. It was bad enough having to endure the elements, but it felt like such a waste of time, having to wait it out until daylight when the chance came for someone to see us. There was more purpose during the day.

I sat up quietly so as not to disturb Dylan, and waited for full daylight to emerge to see what kind of day it would be. The sky was grey and cloudy, and when I looked down the mountain, it was so foggy I couldn't see trees, grass, mountains, or anything in the distance. The fog spread out like a blanket in all directions, so our visibility was limited to just a few feet in front of us. This worried the hell out of me—how cold was the day going to be, or the next night, I wondered, frowning. Feeling the sun on our skin and thawing out was the best part of the morning. To not feel it would be stripping that away, too. It was a daunting feeling. I didn't want Dylan to be cold throughout the day and the night, too. Then something else dawned on me, "How will we see any planes, and how will they see *us* in this fog?"

Dylan stirred and opened his eyes. Clear, peaceful eyes transformed into concerned, sad eyes in a heartbeat. He sat up, commenting on how foggy it was and how the sky was cloudy, "I can't see anything, Mum," he said.

"I know, it's so foggy this morning. Hopefully, it clears up soon," I replied reassuringly, trying to be positive even though riddled with worry and fear.

We stood up, stretching our bodies out and shaking off the discomfort. Dylan, eyeing the sky, said, "Mum, what happens if it rains?"

I thought for a minute before replying, "Well, we will take our clothes off to our underwear and hide them under the huge tree over there," I said, pointing to the enormous tree.

At this moment, I thought our bodies could do with a shower. From my perspective, I felt so dirty and yucky, especially from my period, I would gladly welcome rain. Better still, we would be able to drink it! If it could, my mouth would have watered at the thought.

We talked more about what we would do if it did rain, Dylan saying we could air dry off before getting dressed again, but mostly, we spoke about how amazing it would be to drink the rain. We visualised us standing there, mouths wide open, and how we would let the rain fill them before swallowing and repeating until we could drink no more. The more thought I put into this, the more I was worried about us being wet and cold, or colder, if that was even a thing.

As the morning progressed, the fog began clearing, but the grey clouds stayed, warding off the sun. We sat in our bush lounge room and waited to see what the day would bring. The air was cold, the Mount Royal was quiet. There was no noise at all—everything was quiet and still. Dylan asked if I thought we would still be here at Christmas, and if so, would Santa still come?

That did it for me. I burst into tears because I knew if this were the case, it would mean both of us would be dead. It was a completely overwhelming feeling thinking of my baby dying out here, so young, hungry and dehydrated, or being attacked by an animal and feasted on, or any lot of things that poured into my brain at that moment. I took a deep breath, brought myself back into check, and cast away the worrying thoughts.

I noticed vines that looked like little Christmas trees growing on the ground all around us. I stood up, feeling like I would vomit. I walked over, picked one out of the ground, and said if we were still here at Christmas, we would plant one of these trees in our lounge room. "Just like a Christmas tree. We will collect rocks, and you will name all the gifts you want Santa to bring you. Remember, Santa watches and knows. This way, he'll be able to leave your presents at home, so you don't have to carry them through the bush."

Oh, the things we come up with. Like millions of mums worldwide, I have always told a few lies to allow for the wonderful magic and joy of Santa, the Easter Bunny, and the Tooth Fairy. We both love Christmas—decorating our tree, putting the lights up, buying gifts, and wrapping gifts. To see the way his face lights up when he gets his gifts is priceless. The thought of him not enjoying any more of them was sickening.

With the clouds becoming darker and greyer, I began to think about shelter. I would have to make the roof far thicker than it was, so I started pulling out more shrubs, while Dylan collected more sticks.

We had exhausted the area we were in of usable materials. It was pretty sparse here, and the trees were so tall I couldn't reach the branches to pull the better foliage down to use it. We had to go back down the mountain a little way to access more of what we needed.

Climbing down together, we clung onto the heavier, unbreakable limbs and branches from trees to safely descend the exact way we had first climbed up. Looking back down the mountain, down in the direction we had come, it didn't make sense because nothing was familiar from when we first trekked up. It looked like a completely new section. Again, the bush deceived us, or was it my mind?

My thoughts bounced around all over the place. Do we descend? Do we stay put? Were we safer here? Or were we safer at the shack back down the mountain? "No," I'd argue with myself, "Here is best, where we have a better chance of being seen."

My thoughts flicked around back and forth like a yo-yo. Mostly, my many thoughts were silent, but sometimes, I would share them out loud with Dylan in the hope they would make better sense or process them somehow. Hearing me think out loud, Dylan pointed to the huge tree and said we should go under it and hide there out of the weather. He was the voice of reason, yet again. Sold on the idea, we walked and searched the ground in all directions of our area to find anything we felt useful.

I noticed to the right of our mountain, down the hill a bit, a group of much smaller trees-the size of people. I knew I could reach it easily enough and break some branches, so I explained my plan to Dylan, asking him to sit on the side of this mountain and watch me rather than come with me. I assured him I would only go down a short distance, and we would remain in sight of each other the whole time.

He could see it was a steep slope, and the trees didn't look very strong due to their small size. Given that he could see me, he agreed to the plan and sat down. I slowly started taking one step before the other, descending the slope. I got to the first small tree and broke off a few small branches with all my might. It wasn't much, but it was something.

The ground felt sloppier than it looked. I turned around to ensure I could still see Dylan and was extra cautious. I thought it was probably best to slide down on my bum, using my feet and arms as support, rather than risk losing my footing and slipping down the hill and out of Dylan's sight.

I could only reach about five small trees before it became dangerously sloped. I broke off what I was able to within arm's reach and threw them as close to Dylan as I could. He stood and collected them while I crawled back up the mountain hill. I commented it wasn't much, but hopefully, they would provide enough shelter with the abundance of fresh leaves on the branches, unlike the ones we had, which were relatively thin.

I reached the top where Dylan was sitting and plopped down to catch my breath. I could feel my body begging for rest and fuel. Ignoring my body, I focused on the present task as the grey clouds formed into rain clouds. I felt a slight breeze and noticed the trees beginning to wave. It felt gloomy.

The wind picked up as we worked on the shelter, and the air temperature plummeted. It would have been the perfect weather to be huddled up all snug on the lounge under a warm blanket in our PJs, but here we were out in the thick of it.

Physically, I did not want to move, but I had to. We worked quickly, gathering what we could, and taking it over to the big tree. The top of the fallen tree was so thick and bushy that we couldn't see the sky through any branches or leaves. In the wind, you could hear the large tree branches groan. The strong breeze whipped through our hair as we stacked our sticks, branches, and shrubs firmly in place to block the wind or rain from coming in behind us. Done.

We sat on the ground, shoulder to shoulder, in the small shelter as the wind intensified. I kept a few branches aside that I thought we could use as blankets like we did the first night. Covering our legs with them, the leaves felt cold against our skin. "I know the leaves are cold, but give it a few minutes of them resting on our skin, and they should feel warm from our body heat," I told Dylan.

The wind was picking up even more and was coming straight towards us, hitting us directly in the face. It was still daylight—around 2 p.m. I guessed, but it could have been any time of the day with no sun. Already, we were shivering as the cold wind lashed out at us. I put my arm around Dylan to offer some comfort and, again, was weighted with the guilt of my decision to walk beyond the sight of the car on that first fateful day.

I started to believe that being directly under the tree was making us colder, even though our backs were against it. It felt like we were sitting out in the open with the sound of wind all around us. I needed to rethink this plan; this wasn't going to work. Our teeth were chattering, our bodies were non-stop shivering and beginning to ache, the branches weren't keeping us warm whatsoever, and we hadn't even lost daylight yet.

I wondered if it was best for us to move and walk around since that would help us keep warm, but I was so conflicted because I didn't want Dylan to use up any more energy than he needed to, being so young and little. I needed him to conserve his energy to keep him alive. "Think, think, think," I said to myself.

I knew from previous nights that laying in the dirt felt warm once we had sat or laid in a position long enough. I thought if we headed back to the shelter from last night, we could use the extra branches and shrubs we had collected to protect us from the wind and cold more adequately. I ran through my plan with Dylan, who agreed and began collecting what he could take with us.

Pushing into the wind to walk the 10 to 15 steps to our bush bedroom from the night before was exhausting. I checked the overhead cover I had made to see if the wind had made it unstable or weak. To my surprise, it was holding together quite well. Go me! I would have gone well in those reality survival shows! Pity we didn't have the film crew, medics, and helicopter on standby here, not to mention the prize to justify it all.

With daylight fading, we went to our bush bathroom before preparing for bed. It was cold on an average night out here; we knew we would be up against it tonight. I resigned myself to the fact that despite my best efforts, keeping Dylan warm in this cold wind would be near impossible. All I could do was reassure him, give him lots of cuddles, tell stories to keep his mind off the situation and do my best to get through the night.

After a while, we didn't speak anymore, both of us lost in our thoughts. "Mum? Do you really think we will get out of here, or will we live here forever?"

My heart sunk into the pit of my stomach as I reassured him we would definitely be getting out of here, one way or another.

Instead of the usual stillness and quiet of the night, all we could hear was the whooshing of the trees surrounding us. The sky was completely black. It was darker than usual because there were no stars and no moonlight.

Not having any real shelter or being able to hide from rain, lighting, and wind, if a big storm were to present itself, the only thing we could do was wait it out and hope it passed over us quickly. It would be bittersweet, though; the joy of being able to drink from the rain was excruciatingly attractive. Oh! The feeling of the thought of it was hard to put into words. I wanted it so bad, but at what cost would it come? It was a fuck yes or a fuck no moment.

I thought constantly about my family now, wondering how my other children, Sarah, Daniel, and Tim, were, and if Daniel had got home okay and had contacted Sarah, by now surely worried about where we were. I thought about Mum, Dad, and sister Belinda, and how they were doing, what they would be doing right now, and if they had tried to contact us and were wondering where we were. I would think about my colleagues and wonder what they thought of me not turning up for work. I wondered if they were worried or just thought I'd not bothered and dismissed me as a liability. I would try to guess what was happening in the lives of the people I loved and cared about.

I thought Sarah would for sure know something was wrong. We texted nearly every day, and if Daniel hadn't reached out to her, I thought she would definitely reach out to him, Mum and Dad, or my friends to see if anyone had spoken to me.

Timothy, my second youngest child, would have gone with the flow of his siblings. If they were worried, he would have been, but if they weren't, he would have felt no reason to be alarmed, even though this is out of character for his mum. He wasn't one to worry about much.

I thought of all the scenarios that could possibly unfold. Daniel deciding to stay with his mate, and the alarm not being raised to him getting home and contacting his siblings and then them all meeting at our house, trying to piece together where the hell we were. The latter worried me because I was sure they would have been beside themselves with concern, but at the same time, it meant we had a hope of being rescued.

Come Friday, which was yesterday, campers or visitors or a ranger, someone, would surely see our car still parked there, notice it had been there for a while, and possibly report it. I hoped.

The stories and possibilities ran through my head like a slide show of movies. I believed wholeheartedly in at least one coming true and leading to our rescue.

With the branches over our legs, we pressed up hard against each other. I kept my arm around Dylan and held onto him tightly. I dozed, he dozed, we'd wake, talk, sit, huddle, try to sleep, shiver, and shake.

As the night progressed, the wind howled, and the branches and trees around us were pushed and pulled in all directions by the force of the gale. I was hopeful it would rain, but not until daylight came, so we wouldn't be even colder than we already were. I continued to doze off and on. Most of the time, it felt like my eyes were only closed for seconds before springing back open.

I woke to the feeling of cold drops of water hitting my face and forehead. I could hear and feel the wind had become much stronger, but all I could think about was that it was raining. Rain! OMG! Water! Dylan woke up seconds after I did, "Mum, is it raining?" he asked excitedly.

"Yes! I think it's starting to!" I replied, as excited as he was. "If it does, I want you to stand up so only the top of your shoulders gets wet, tilt your face to the sky, and open your mouth as wide as you can so the raindrops go in, ok?"

Nodding enthusiastically, he said, "Ok, I will."

The raindrops were getting more regular, so we moved into a space without treetops between us and the sky, leaned our heads back, and opened our mouths wide to catch the raindrops. We couldn't believe it, just our luck; not one, yes, not one drop went into our mouth. A raindrop or two fell on our forehead, cheeks, the top of our heads, and everywhere else except into our mouths—a cruel twist of fate. Our necks began to ache from the strain, desperate to receive at least a few drops on our tongues.

We really needed this. "Come on! For fuck's sake!" I yelled in my head silently. I thought as long as Dylan could get something, anything, then I'd be happy. I wasn't sure how much longer he could survive without water, but I knew it wasn't another week, that's for sure. I pushed the thought out of my mind. I couldn't entertain the idea of him being out here for another day, let alone another week! As for me, I was surviving for him. I had to keep going and push through to get him out of here.

I didn't feel pain in my body; it was in my heart. I already secretly decided that if Dylan, for whatever reason, didn't make it out alive, then I wasn't going to either. I knew that the moment Dylan died, I would use a rock, stick, or anything I could find to cut my wrists and bleed out beside where he lay so he would never be alone, and we would always be together. If we were ever discovered, at least we would be found together. In my mind, it would be both of us or neither of us.

We stood in the middle of the opening to the sky, wrapped our arms around each other, and cried dry tears. Shivering in the cold wind, we felt helpless as the few drops landed on our bare skin.

After a few moments, we unlatched our arms and walked back, side by side, keeping close to each other in the dark, to our bush bedroom before, lying down. We lay listening to the swooshing of the trees, protesting against the cold wind. I really understood how they felt. Trying to get to sleep, I angrily thought, why have a sky so dark and covered with rain clouds if it didn't bother raining properly? Only the Mount Royal could answer that, and even then, I decided it would probably be a deceptive answer.

Eventually, we got back to sleep, for how long, I don't know, but we woke with a start as a wild gust of wind tore through the bushland. It was quite literally howling. I remembered hearing the same thing at home and the wind battering the house, insistent on finding a way in through any crack or crevice it could find. I remember thinking how lucky we were to be safe and secure inside rather than out.

Dylan woke up, and I could barely hear him say how windy it was from the noise of it swirling around us. I put my arms around him, leaning close to his ear and saying, "Don't worry, we're going to be okay," through chattering teeth. God, it was cold!

I directed him to sit on the ground before me to shield him from the wind, dirt, and foliage swirling around us. "Keep your head inside your shirt!" I told him loudly above the howling. I leant over, trying to block some of the wind with my body, but it was coming at us from all directions, so it was pretty useless.

It was coming in waves, full force one minute, slower the next. I could hear when we were about to cop the powerful gust, because you could hear it coming. We braced ourselves as it bounced angrily off the mountains like someone on a rampage zeroing in on us. It was like the gods were saying, "Who are you, and why the hell are you here?! You shouldn't be here!"

I felt like replying, "Show us the way, and we'll be happy to go!"

After a wave passed us and the calm between them came, Dylan would poke his head slowly out of his shirt like a turtle coming in and out of its shell and stand to stretch out before crouching down again and ducking his head back in his shirt. Thankfully, the more powerful lashes of wind only lasted for a few seconds before leaving our mountain and moving on to the next, doing its rounds and trying its best to intimidate anything in its way.

It seemed to go on for hours with no signs of letting up. My legs were sore from the constant standing over the top of Dylan to protect him, and I was tired from lack of sleep. We were both so exhausted and dehydrated. I told Dylan to lie down and covered him with the branches. Then I crawled under them next to him, sat down, and wrapped my arms around him while he dozed back off to sleep.

With my arms bare and bearing the brunt of the wind, I eventually laid down wholly spent and listened to the roar of the gale circling us. I can't remember dozing off, but I was happy when I opened my eyes, and it was daylight.

Chapter 10

START LICKING

Day 7 SUNDAY

I felt relieved we had survived the night, and the wind had died down to a gentle breeze that calmly brushed through the trees, creating no more than a slight movement among their branches. The grey clouds were still in the sky, but there was no sign of rain. I sat there asking myself, why won't it rain? I have never been a religious person, praying to a god or the heavens above. Still, I began asking a higher power, ANY higher power, begging them or it for some rain, some water, ANY water, either just for Dylan or for us both.

I sat quietly, trying not to move so Dylan could get as much sleep as possible. Looking down the hill, it was quite foggy again. The air was brisk, and I knew it would be another cold and cloudy day. Thinking about the night before and how windy it had become, I dreaded what would come if it were a carbon copy of yesterday's weather. I stared ahead of me at the bush. I had really begun to hate the Mount Royal. It was entirely out of character for me to hate anything or anyone. I always looked on the positive side of things, but it was wearing thin and testing me in all areas—my mind, body, and soul.

I looked without seeing the bush, just gazing off ahead of me when my eyes were drawn to a bush of shrubs. They weren't special in any way, just shrubs, the same we had seen a million times over the past seven days. "Wow," I thought, "We've been here for a whole week." I looked at the long strands of grass growing from the bottom of the shrub and noticed drops of water on them.

I instantly stood up and walked the 3 or 4 feet to them, bent and gently so as not to disturb the drops of water, took one in my hand, lowered my mouth down, and started to lick the strand. I couldn't believe it! The feel of that moisture from those drops on my lips was beyond amazing.

I heard Dylan say, "Mum, what are you doing?"

I looked up at him and smiled. "Baby, come here!"

He got up with a puzzled look and came over to me.

"Look at these plants—they are wet! I want you to carefully bend down and lick all the moisture you can from them," I directed him. "Watch out for the sharp, prickly points. I will hold them for you so you can lick as much as possible."

He was so desperate for water that I didn't need to repeat it.

He dived straight in, but carefully as I had instructed. He quickly got the hang of it and began holding the strands himself.

I bent down and started licking the strands on the opposite side of the shrub as well.

Feeling a few minor paper cuts on my tongue, I again warned Dylan to be careful. He didn't care about getting a few cuts if it meant he could get the moisture onto his lips. He looked like he was in pure heaven, and the joy of seeing him finally take in some water was the best feeling I had all week.

Between us, we licked dry all the strands directly around us. The sun still hadn't emerged, but it was light enough to walk around safely and look for more shrubs. We walked over to what we called our garden area, and my eyes lit up like a Christmas tree. There were tiny drops of moisture everywhere on the leaves of the small trees, and I spotted some fallen leaves that had a few drops on them as well. It was like finding tiny specks of gold. Coming across any water out here was something rare, rich, and valuable. I was so excited; I didn't know where to begin!

I left the leaves with the bigger drops on them for Dylan and very slowly and delicately picked up the leaves to carefully tip the drops into his mouth. Already, I could see his lips fill with moisture, and the cracks and dryness were disappearing. He frantically drank the moisture, licking the leaves dry and quickly moving on to the next.

He was so excited and desperate to feel the moisture on his lips that he didn't stop for what felt like hours. It felt so good— like we were children and parents running around licking and tasting all the edible plants and treats they could find in Willy Wonka's Chocolate Factory.

Once we had licked every leaf on every tree and all the long grassy shrubs we could reach or see, we-sat back down in our bush lounge room. Even though we had effectively had a big drink of moisture, we felt exhausted from the effort, but so happy that we couldn't stop sharing and expressing how good it felt.

Sitting there, I felt pretty proud of myself for being so resourceful—it was a proud and relieving moment for me!

Our mood was cheerful, but the sky was still gloomy with grey clouds and no sun. I absently said I didn't think there would be any planes or helicopters today. Dylan looked at me. From the look on his face and the tone of his answer, I immediately picked up he was disappointed, so I redirected my comment and observation to keep his hopes and spirits up. "But you never know; people might like to see the bush when it's grey and cloudy, so the sun isn't in their eyes. I better prepare myself to jump and wave like crazy, just in case."

His face lit up, "Ok, Mum, I will help!" he said, smiling and enthused.

I reached out and touched his face, saying, "I know you will."

With nothing else to do but wait and listen for any planes flying above, I suggested he lay back and sleep. Although a little brighter, he still looked wiped out from last night's lack of rest. The best part of an overcast day was that there were no freeloading, buzzing, or biting fucking flies! Bliss!

Different weather meant different elements, exposure, and habitats depending on where we were. For example, lower down on the mountain, we could hear all different bird sounds; when we were up high, we hardly heard any bird sounds. With more exposure up high, we also had more sun to stay warmer throughout the day. I also noticed far fewer animal droppings on the ground here, but down the bottom, there were heaps. The trees were much taller, and not so many here, while down low, they were thicker, and there were more of them.

I sat thinking while Dylan slept. As usual, I looked at him, watching him dream, and wondered how I would get him out. I was concerned about his wellbeing even after we got out, and worried about my other children. My thoughts went dark, thinking what would happen if I died out here. I wouldn't see my children grow up and see what they made of their lives or be there to support them. In many ways, I mainly did everything for them, even while married to their dads.

I was a young mum when I had Daniel—only 16 years old. His dad was much older and lived life his way—travelling a lot, hardly ever home. We didn't stay together for long, and when we separated, he decided he didn't want to be in Daniel's life. It was devastating for both Daniel and me.

The marriage to Timothy and Sarah's dad lasted for 14 years. He had minimal contact with them now due to being dominating and abusive. It was sad for the kids. He had no regard for other people's feelings or respect for himself or others.

Dylan's dad and I were together for five years. We separated when Dylan was still little. He struggled to comprehend the meaning of parenting, deciding he would pick and choose when he wanted to be in Dylan's life, which was generally when he was single and not preoccupied.

The children, I felt, knew they could rely on and count on me to be there for them all through their lives up to this point. It never occurred to me it would be any different. When they were struggling or needed a hug, I was there. I was there through every experience, whether it was sad or happy. It pained me greatly to think I'd never see them married and having their own children and not getting to be a hands-on grandmother. It was all hard to think of. I couldn't imagine not ever seeing my brother, sister, or my parents again. Our family gatherings were wonderful, especially at Christmas—putting up the lights, decorating the tree, wrapping presents, and watching Christmas movies like my favourites Home Alone 1 and 2. Dylan would never experience anything of life—he was so young and had barely begun his life. What if he never got to taste food again, have another birthday or Christmas, or never saw our pet rabbit Bunnygirl, whom he loved so much again?

Realising it could all be possible hit me like a ton of bricks. Every thought came rushing at me like a tidal wave. I closed my eyes—hurting, grieving already for what I could be losing. I began silently saying a thank you to each of my loved ones, telling them I loved them, and saying goodbye. It ripped me apart. I was crying but was still so dehydrated that the tears were dry. I had to open my eyes; it was too emotional and became too much to think about.

I sat up and thought, "No, this isn't it. We WILL get out of here."

Spurring myself on, I forced myself to think positively. "It's going to rain," I said with determination, thinking about the drops on the leaves and catching them. I decided when it did, I wouldn't wake Dylan up until it was enough to get excited about. I wondered if I could use a stick and dig a hole in the ground to capture some water. Remembering back to the leaves from the morning, I thought I could place a whole stack of leaves into the hole at the bottom to catch and hold the water.

Feeling focused on this next mission, I felt more at ease and relaxed somewhat. I looked at the landscape and thought no one in their entire lifetime could count how many trees there were in this place. There were so many for miles and miles. Even in the other national parks we had walked, I had never seen so much bush in all my life. I wondered what time it was, maybe 11 a.m.? 12 p.m.? I wasn't sure, with no sun to guide me.

Dylan had awoken and sat straight up. "Mum, what would you eat right now?" he said with some energy.

I thought maybe he had a food dream or had become so hungry that it was all he could think about. "Anything. Even pumpkin," I said with a laugh.

Dylan didn't know what a pumpkin was since I didn't buy it. He'd only eaten it when he was a baby, so didn't remember it, and I had to explain to him that it was a vegetable, one I didn't like.

He was surprised because I loved my vegetables but couldn't stand pumpkin or Brussels sprouts. Right now, if I saw either, I would hoe into them without hesitation. It's incredible what you're willing to eat when desperate. "What would you eat?" I asked him.

"KFC."

I laughed. We did not buy takeaway often—traditionally, only on birthdays. On his last birthday, the month before, he had chosen it and gotten food poisoning, so I was shocked he'd say he would eat it again, but hey, like I said, when you're hungry enough, you'll eat anything.

We talked about food, drooling, imagining eating KFC or our favourite…Mango smoothies. "Mmmmmm, yum!" we said in unison at the thought of tasting it.

I told Dylan about my idea of the hole and leaves, and he thought it was a good idea, "Yeah! I'll help!"

I wanted him to, but he said, "I know you like to help, but with you not having any food or much water, you need to save your energy and strength."

I could see he was a bit disappointed in this, so I said, "Ok, how about you collect as many leaves as you can, and when I have finished digging some holes, you can put them in the bottom and line the holes with them?"

Smiling, he said, "Ok!" happy to be needed and helpful.

I stood up and said, "OK, let's get started."

We walked over to the dirt patch where no tree branches were overhead, and I got a stick and started digging for China. The dirt was not soft, but it wasn't rock hard either. I was able, with a bit of force, to loosen the dirt and begin to start a small hole. I kept at it, and before long, I had chiselled out a hole the size of a bowl around 20 cm in diameter.

"How about I dig a few of these around? We should be able to get some water in one or a few of them?"

Dylan thought that was a good idea, so I kept digging, but not too close to our bush bedroom or where we would walk around at night. I didn't want us tripping or stepping awkwardly in them in the dark and breaking an ankle or something.

Once I was done, he would come along and line them with a few layers of leaves, leaving a hole deep enough to catch the water.

After digging out the smaller holes, I thought I'd dig a bigger one so we could wash our faces and hopefully even take turns sitting in it and bathing. At the very least, we would have plenty of water to drink. That's if it ever bloody rained! What a luxury it would be! I loved having baths at home on rare occasions, although finding the time was always challenging. I rose early and was always on the go until bedtime. I never was into self-care like I should have been. It took a lot of time to realise the importance of this in later years.

I told Dylan my idea about the bath, and he was keen on the idea, so I got digging, and he set about collecting more leaves. Gouging out the earth, I began to feel a bit lightheaded and started to heave. With nothing in my belly to bring up, I kept dry heaving.

I visualised the water in the holes and saw Dylan drinking loads of it to take my mind off feeling sick and to motivate me to keep going.

Dylan's brows were crossed as he looked at me with concern.

"I'm ok," I assured him, continuing to dig. The heaving continued for another few minutes, and Dylan asked why it was happening. I didn't know for sure, "Maybe because I have nothing in my stomach, and the digging is making me hungrier," I replied.

"Mum, I'm hungry too, so why am I not doing it?"

I replied, "Because you're not digging, it's a good thing I asked you not to so your body doesn't feel sick. The last thing I want is for you to throw up any water you have in your body."

I continued digging and turned my head away from Dylan when I heaved so he didn't get upset by it. While I was digging, he collected some rocks and other sticks buried in the dirt to keep occupied. I made it into a kind of treasure-hunting game. "We can put these around the holes to help keep the water in," I said, as I kept digging.

Once the hole had become about half an arm's length deep and approx. 40 cm in diameter, I stopped and said, "That should do it." Physically, I didn't think I could have dug anymore anyway. I'd been at it for a good few hours. I looked up at the sky and saw low grey clouds. A good indication that rain was on the way. Fingers crossed.

"Ok, baby, put the leaves in the bottom, and I also think we should break some small branches to use for the inside of the hole to help the water in better too," I said, not entirely convinced it would work, or if we'd even get enough rain to fill them. I hoped and visualised the holes full of water and drinking and washing ourselves with it. It was worth the shot.

After lining the holes with the sticks and rocks, I suggested we lay down in our bush bedroom to get some rest before nightfall in case the weather worsened. Standing up, I felt another gush of blood exit. I was too exhausted even to do a thing about it. There wasn't much I could do anyway. I couldn't start hacking into what was left of my T-shirt; I needed it to keep my body safe from the elements. I knew blood was exposed to the crutch and butt areas of my jeans, but I had already explained to Dylan what it was—a part of me and a part of life. He got it and didn't care. It was what it was.

In silence, we both sat on the ground for a few moments. We didn't speak, lost in our own worlds for a while. I wondered when and if we would ever be seen and rescued. Dylan broke the silence to ask, "If we were at home now, what would we be doing?"

Thinking back to our regular routine, I said, "Well, to guess the time, I would say I would not be far from getting home from work, eager to see you and ask how your day has been. Then we would chat about dinner, deciding what to make from the meat I would have gotten out of the freezer before work."

He said, "Yum, dinner."

I sighed, "I know, let's pretend we are making Bolognese," to which we both said, "Oh, yummmmm!" in unison.

We visualised it for a few moments, and I dreamed of being nice and warm in our home, cooking and dishing out the pasta and Bolognese minced sauce into our bowls and sprinkling the grated cheese over the top.

Dylan suggested, "Let's talk about going out to dinner."

I said, "Ok, let's pretend we are at a restaurant and about to order our meal. I will ask for water to be brought to the table and ask you what you are ordering."

He said, "Everything!" and we both laughed.

"I'll have the same!" and we laughed some more.

Dylan asked, "Can you order everything on the menu?"

I said, "Sure, if you have lots of money to pay for it, then yeah, you can order what you like."

He asked, "Can we do that one day?"

I laughed and said, "I don't think a restaurant table would be big enough to fit all the food from the menu, and we would feel so fat and sick from trying to eat it all."

We laughed again, then fell into another silence, dreaming about it anyway.

Dylan asked what I thought his brothers and sister would be doing, and I said I thought they would be at our house, all trying to figure out where we were. I remembered the piece of paper I had scribbled down the places we had visited the night before we became lost. I had fortunately added the Mount Royal to the bottom of it. I couldn't remember if I had left it on the table or thrown it out in the bin that morning before we left, but Dylan said I had definitely left it on the bench.

I said, "Are you sure?" he said yes, adamant I had left it on the bench. I sure hoped I had. It was something I had thought about multiple times.

I put my arm around Dylan and said, "Come on, let's get some sleep while we can."

Within minutes, we were both fast asleep. I don't know how long I had been asleep-when I woke up with a start and an uneasy feeling. I had a nagging feeling about staying up here on the higher ground and began to feel panicky, wondering if we should continue to stay here.

I thought about the disgusting brown water puddle we had encountered days before. I wondered if we could try and cut through the bush by going down the mountain where the puddle was, without backtracking the way we had come.

I turned my head and looked at Dylan. As he lay still sleeping on the ground, I put my hand on his side and whispered, "I will get you out. I promise."

While looking at him, I noticed how much weight he had lost already. His bright red shirt seemed a lot bigger than it had that first day.

It suddenly occurred to me that when the sun was in the sky and we heard the next plane, we could put his shirt on the longest branch we could find and wave it around, like a flag. Maybe that would attract someone's attention? Almost all of the planes that flew over us came in the direction behind us. If they had come in front of us, there was more chance of them spotting us due to the tree formation here.

In my mind, I was already playing the moment out in my head, like a scene in a movie. We could hear the plane approaching us, and then Dylan quickly rushed to remove his shirt. I would position it between the branches, hoist it as far as possible into the air, and wave it around to signal the plane. I looked at Dylan, excited about my plan, and couldn't wait to share it with him. I hoped it would work. I missed my other children, friends, and family a lot and was desperate to get Dylan out of here and see them all.

It felt like hours went by before Dylan woke up. I removed my arm from where he had been resting on it, and he opened his eyes, smiled, and said, "Hi, Mum."

The love I have for this boy of mine is immense. I looked lovingly back at him and said, "Hey, baby. That was a good sleep you had."

Nodding, he asked, "Do you think we'll see any planes today?

I told him I sure hoped so and explained my idea to him.

"Yes, that's a good idea. Let's try that, Mum," he said excitedly.

We laughed about what people would do when they saw a red shirt waving around at them. "They'll be like, WTF is that?!" I said while we laughed more about it.

I stood up. My ass was so sore from sitting on the ground for so long that it was beyond numb. As soon as I moved, it made it clear to me that it wasn't happy about being in the one position on the hard Earth for as long as it had been. I didn't bother thinking about my periods or changing the cloth anymore. I had accepted it for what it was.

"Do you want to rest while I find a big stick to use for the flag?" I asked.

"No, I want to help, Mum," he said, getting to his feet and stretching.

"Where do you want to look first? I asked.

He pointed to a really large tree and said, "There."

I nodded, "OK then, good idea."

We walked over to the tree, scouring the ground in all directions for a branch the big tree may have dropped, but we could only find smaller ones, nothing of use.

We walked towards the left side of the mountain and came across a decently long branch, which we agreed would do the trick. I leaned over and picked it up. It was long and heavy but not so thick that I couldn't get my hands around it to keep it in the air while waving it around.

"Perfect," I said.

"Mum, are you sure you'll be able to lift it when a plane comes?"

I smiled, "You just watch me."

He smiled back and took one end of the branch, and we walked back to our bush lounge room, which was the spot that had the clearest vantage point of being seen. We placed it on the ground, then sat down and waited and waited.

Sometimes, we'd sit in silence, but mostly, we talked about anything and nothing. There wasn't a lot of daylight left, and the worry we wouldn't see a plane fly over started to become real and dulled the excitement of our plan. I couldn't believe it; not one plane had passed over. In fact, we didn't even hear any. Nothing.

"We might have to wait until the morning, Dylan," trying to sound optimistic to hide my disappointment. "We also might have to scale back down the mountain to that muddy puddle we saw if it doesn't rain tonight," I told him, leaving my fears about his survival if he didn't get more water soon unspoken.

"That looked so gross, Mum, and you were nearly sick when you tried it," he recalled.

"Yeah, true, it was worse than the urine," I laughed, but I thought it was better than nothing. I was beginning to doubt we would be rescued from our position, having not heard a plane all day. I couldn't shake the feeling we should move. It didn't look like a rescue would occur if we stayed here longer. I started to stand up.

Dylan asked, "What are you doing, Mum?"

I explained what was on my mind, "I think we need to move from where we are."

He sat on the ground with an exhausted look on his extremely dehydrated face. "Really?" he asked.

"Yes. Come on, let's go," I said before either of us could think or hesitate about it any further. I knew Dylan did not want to move; I felt the same, but I was sure we had to.

Getting both him and me motivated, I asked, "What if we try to go down the mountain from this side—the left side?"

He asked, "Why do we need to leave from this spot?"

I explained to him that I thought we would be too small to spot here and needed to find water anyway.

I began walking towards the edge of the mountain in front of our bush lounge room and towards the left side, facing the direction we had come up from. I walked back and forth along the mountain edge, trying to calculate our route. No matter where we started, it was steep. I knew the section I had climbed had flimsy trees that couldn't support us.

Looking further to the left, I saw a large tree log on the ground and pointed it out to Dylan. "We need to get to that tree log, and then I'll work out the best way down," I said.

"OK," he replied, not excited about it.

"We can do this," I said convincingly while looking down at the steep decline. "We will need to sit on our butts and use our feet and hands to crab down slowly. Stay right behind me, and if you or I begin to slip, stop straight away, ok?" I instructed him.

"Ok," he responded.

Focusing on the log we needed to get to, I checked in constantly to ensure Dylan was ok. As we got closer to the log we got, the bushland began to change. It was really thick here, and I could tell it would be a task to reach the flat area I could see.

We zigzagged around dangerous steep parts that would have been impossible for us to climb down as they had long drops with no way to get down safely. Instead, it meant us having to navigate the thick bush skirting the drops. Reaching the tree log was like an obstacle course, but we finally did.

It was nothing like it had looked from the top. It was much larger up close; the length of it was about the size of six cars lined up boot to bonnet. We walked alongside it, gingerly watching our footing and stepping over its fallen branches and the shrubs wildly growing on either side.

I looked straight ahead and was still unable to see the bottom of the mountain. It was so steep and thick with more shrubs and trees. I could see what looked like a bit of a lip up ahead, but I couldn't judge if we could get down from the lip to go further down the mountain or not, and I couldn't explore it because of the immense drop-off. The right side of the tree was so dense I couldn't calculate what was on the other side of it, or even if we could push our way through it. I was running out of options and ways to continue down the mountain.

I looked back to where we had climbed down and could barely see the top of where we had started. We crept slowly, holding onto the log for support and inching forward until we got to the end of the tree log when we saw yet another. This one was half buried in the ground alongside the wall of the mountain that ran along the edge of the giant drop-off.

There was a trench of dirt where the mountain met the log that was just wide enough for our butts to sit, but our legs would need to dangle over the tree log that ended where the drop-off started. I didn't know how far the drop went, but I wasn't about to risk leaning over to find out. It was precarious enough. The trench wasn't an ample space, but it would do for a few minutes while I figured out what to do next.

The view was spectacular from here, so green and lush—picture-perfect. We both looked at it with amazement and couldn't believe how nice it looked. With the mountain drop-off and so many trees and shrubs, there was no way we could stay here if we wanted to be spotted, and there was no chance of finding water either, so we would have to move, but we could huddle there for the night if we laid head to toe along the trench. It wasn't wide enough to lay safely side by side.

We pretended we were on a giant swing, and I said it looked like a massive secret garden, while Dylan said he thought it reminded him of a kind of tunnel with a view. We always used our imagination to bring some joy to a moment and distract ourselves from a situation. While we talked about how it reminded us of Jurassic Park, I continued thinking of what to do and kept an eye on the clouds.

I thought if it did begin to rain, we could climb back to our water holes. The look on Dylan's face told me he was far from impressed about having to climb back up again. "It's going to be dark soon," I said. Being in this location, it was getting darker far quicker than if we were in a clearing in the bush. "How about I break off some branches from this tree and the ones growing beside it to help us keep warm, and we lay here for the night?" He was much happier with that idea than climbing back up the mountain again.

I told him to stay put because I didn't want him moving around without me and risking falling down the drop-off. "OK, Mum. You be careful."

I assured him I would while standing up again and taking a few steps back so my back was touching the mountain wall. I took hold of the big log to work along it to the branches, and continued chatting with Dylan so he wasn't nervous about me moving away from him. Collecting a few, I asked, "Do you think we have enough?"

With a confirmation from him, I held the branches under one arm, and used my free arm to hold onto the log and return to him.

I sat in the dirt trench beside Dylan, gave him three of the bushiest branches, and kept two for myself. I sat and patted my lap for him to lay his head down and rest on. "I'll put the branches over you, and you try to sleep," I said caringly.

"What about you, Mum?" he asked, concerned.

"Don't worry, I'll cover myself with my branches and try to get some sleep too."

As daylight left and Dylan dozed off to sleep, I positioned myself so I was lying down, too, to be a bit more comfortable and warmer. I sat there listening to the silence, and every now and again, I heard a bird sing out. As the night progressed, sounds changed, and I could hear the rustling of movement in the bush. I wondered what and how big it was. Your mind goes to strange and scary places sitting in the dark in bushland. In a way, I was glad I couldn't see what it was. As long as it didn't come closer to us, I knew we would be fine.

I watched the landscape, keeping an eye out. I thought it looked like the scene in Jurassic Park 1 where the two children and the scientist were high up in the tree for protection, surrounded by bushes and dinosaurs, and they woke up to a leaf-eating dinosaur. We weren't up a tree exactly, but the shadows playing on the trees looked like giant dinosaurs were roaming around.

As it so often did, my mind drifted to my other kids, my parents, sister, and brother. My brother was born handicapped with brain damage, so at least he would be blissfully unaware of our situation, but I couldn't imagine what my kids, parents, and sister were going through. Dylan began to stir, cutting off my thoughts, but settled again. "It is so quiet here," I thought, drifting to sleep.

I felt movement and was immediately alert. It was Dylan shifting around. He was uncomfortable, I was uncomfortable, and we both needed to change positions. As we tried to get some movement into our bodies, we noticed how bright the moon was and how much it lit up the area in front of us. It was breathtaking and looked like something straight out of a movie, the way it shone on the trees. "Wow," Dylan commented.

"It's so lovely," I said.

"Yeah, Mum, it is," he replied while we shuffled around for a more comfortable position.

We had gotten so used to shivering, aching, and feeling uncomfortable every night. It felt almost normal. Tonight was different; it was cold, and we were still shivering, yes, but it wasn't freezing like it had been on all the other nights, which was a nice change.

I eventually fell asleep thinking about how and when we would get out, and I ended up having one of the best sleeps since I had been out here.

Chapter 11

DYLAN TURNED PURPLE

Day 8 MONDAY

I woke up around 5:30 a.m., at a guess, in a sea of fog. I could see the sun was rising and was happy that it didn't rain last night, so we couldn't have missed the possibility of drinking from our water holes. I was also pleased that we would have sun today and be warmer throughout the day.

I looked around in all directions as light pierced through the fog. Through the haze, it was difficult to see further down the mountain; the shrub was so thick, and I wondered if it would become even thicker further on if we continued. What if we got stuck or needed help to get back up to this area or back to where we were?

I decided it would be far too dangerous to risk going further, and I needed to forget about the dirty puddle—it just wouldn't be worth it. Instead, we would head back up the mountain to the spot where we started and figure it out from there.

The sun rose higher, shining on Dylan's peaceful, sleeping face. I had gotten to the point that the first thing I did in the morning was to check his chest, which was still moving quickly, and he was still breathing. I had no idea how long he would survive without food and water. I couldn't remember ever hearing Bear Grylls say how long a child could survive without food or water. This can NOT be my baby's final resting place.

I became emotional thinking about it and dry-cried for a few moments, saying over and over in my head, "I'm so sorry I got you into this situation." I tried hard to shift my thinking, pull myself out of the dark thoughts, and pep myself up before he woke. It was hard, though, to see the steep mountains and know what we had already trekked through. I was beginning to feel there was no hope of being able to find a way out.

As I sat there deep in thought, looking down at the log I called the ledge, I saw one small ant crawling along the log. It didn't have a specific colour like a green ant or any other features. I wondered if I could eat it. I picked it up without hesitation and quickly put it in my mouth. It was so small that it wasn't even in my mouth, like a grain of sand. I thought, "OK, if I don't have any reaction to eating the ant, maybe I can find some for Dylan to eat."

After a while, and without a reaction, I looked around for more ants. Unbelievably, there was none. Not even one.

The trees blocked most of the direct sun and sky from where we were, but I could tell it would be a clear blue day. I couldn't see a single cloud. I thought the planes would be back out flying for sure, so we needed to get going again as soon as possible so we could wave one down.

Dylan began to stir and stretch around what I thought was 7 a.m. or 7.30 a.m. The fog was still visible but not as thick as when I woke up. I asked him how he slept, which I normally didn't do, knowing exactly the answer. "Good," he replied.

"Me too!" I said.

He said he was warmer, and that made it easier. I was relieved he felt rested and asked if he was ready to climb again because it was a clear day, and we would have a good chance of flagging down the planes today.

I was starving and desperately needed a drink. I could see Dylan's lips were dry and very cracked again. I told him there was no way to make it down to the dirty puddle—unfortunately, there was no water yet for us, but hopefully, there was a good chance of being spotted and rescued instead.

The look on his face was enough to tell me he really did not want to do any more climbing. After some more talking, he understood and knew our best chance today would be to position ourselves well for a plane to see us. "We can do this. Let's wait a little bit longer so you can wake up a bit more, but the sooner we get up there, the more chance we have of getting our flag up and waving." I was pretty confident today was the day! Well, I was hoping, anyway.

We talked for a while about breakfast and what we would eat if we could have anything right then. I said bacon and eggs, and Dylan was like, "Yes, yum!"

The expression on both our faces said it all. I could close my eyes, taste, and visualise the feeling of putting the fork into my mouth. The idea of crunchy bacon with a small amount of runny egg yolk sending my taste buds into heaven was almost too much of a sensation overload. A far cry from the ant I'd just tasted, that's for sure!

"OK, baby, let's get going. I'll direct you and point out what to hold onto. You'll go first, and I will stay right behind you. And remember, any time you need to stop, we will," I said.

He began to stand up reluctantly and, in a soft voice, said, "OK, let's go."

I also stood up, and we turned around and faced the big log. "Put one foot on the ledge of the tree and use the mountain wall to hold onto, so you are in front of me," I explained, positioning him so I was behind him and could be the backstop if anything went wrong.

Even though we were going the same way as when we went down, it was far more effort going uphill than down. We needed to use more energy and more muscle to pull ourselves up and climb. We continued until we came to the beginning of the tree log, where we rested for a few minutes. Both of us were out of breath and panting, and our mouths were bone dry.

I focused on the direction we needed to climb, so we didn't stray off track. "When you are ready, I want you to take a few steps to that tree, then that next tree, and hold on," I instructed him. "Don't stand straight up as the mountain is going downhill, so you must keep facing and lowering your body towards the ground as best you can." He already knew what to do; I just needed to keep him focused and be sure he knew why.

We continued the zigzag back up the mountain until we reached the point where I said he could get on his hands and knees and crawl the rest of the way. I felt it was safer, and not having many strong enough trees to hold onto, it was better to crawl than risk holding onto a branch that could break and end up with us both tumbling down the steep incline. "We're nearly there. Not much further to go now," I kept encouraging him.

Before I knew it, we were back up to the top and gratefully fell on our backs, exhausted, letting out a sigh of relief. "Thank fuck for that," I thought as we both tried to catch our breath.

Suddenly, I could hear a plane. I sat bolt upright and said, "Dylan! Can you hear that?!"

He sat up fast, too and said, "Yes!"

Forgetting our exhaustion, we got to our feet and scanned the sky to determine which direction it was coming from. Then, right in plain view, we could see a small white plane coming from the direction behind us. We both started waving our arms in the air, jumping up and down like crazy, and yelling, "We are here!"

No matter how hard we waved, the plane continued on. We jumped, waved, and yelled until it was out of sight. Dylan asked, "Do you think it saw us?"

I told him I hoped so. "I hope they're on the radio right now saying they spotted us, and soon a rescue helicopter will be dispatched and drop us food and water." That thought made us both imagine it happening and provided some comfort and hope that it would play out.

We both sat back down, a little disappointed the plane had gone and didn't do another flyover. The sun was at its highest in the sky, so I guessed it was around noon. I often commented on the time to Dylan and taught him how to tell the time by the sun's position so he could get used to gauging how much daylight was left.

Looking around me at the leaves on the tree, I noticed how green and lush they were. I thought they looked tasty. "Gee, I must be hungry," I thought. I asked Dylan if he remembered Bear Grylls demonstrating how he could eat plants found in nature.

Dylan said he did, but also remembered that some made you sick if you ate them.

"Yeah, I'm sure it was berries and leaves with distinct markings on them. These leaves are bright and green with no berries or markings," I observed.

Dylan looked at me with a confused look on his face. "Are we going to eat leaves?" he asked.

The way he said it made me laugh. "Well, I'm thinking about it, so yeah."

He replied, "What if they make us sick?"

I said we could pick ones with no markings or berries and give it a go. I hadn't seen any trees or branches growing berries, so I hoped it would be ok.

As per my rule, I tried anything we did that was new first. If there was no reaction, I knew Dylan would be fine trying it. I stood up, walked 3 feet to a small tree with bright green leaves and no markings, berries, or any distinctive signs that it might be harmful to ingest, and plucked a leaf off the tree. I looked at it, then at Dylan, and said, "Well, ok, here goes!"

I folded it up so it was small, popped it in my mouth, and started chewing. He looked at me with a look that said clearly, "No way I'm doing that!" It tasted awful, but I was desperate for him to eat something, so I reassured him it was ok.

I sat back on the ground next to him, and raising my hand with another folded leaf, I dropped it into his mouth. He took a few chews on it and quickly spat it out. I laughed. His reaction was priceless. I could just tell what he was thinking: firstly, it tasted terrible, and secondly, people aren't meant to eat leaves from a tree. I had to agree with him and spat mine out, too.

My mouth was so dry, and there was no moisture in the leaf, which made it hard to chew. Plus, I was worried I'd choke on it or get a piece lodged in my windpipe. With no water to wash it down, I thought better of the leaf-eating idea. "I'm glad you didn't eat it, Mum," Dylan said, more than a bit relieved I had given up on it. I knew it probably wasn't the most brilliant idea I'd had, but I was getting desperate to relieve our hunger.

We started talking about food and drink again and his experience when he got food poisoning from KFC for his birthday—what a horrible experience that was. The cramping, continuous vomiting, argh! Getting off the topic, we moved on to many other topics, one running into another, passing the time, and distracting ourselves from being hungry and thirsty.

Dylan asked, "What about school?"

I replied, "What about it?"

"What will my teachers and friends think?"

I said, "Well, when you get home, you can tell them all about this, and I'm sure you'll find out what they think."

He laughed, and we talked about how shocked they would be and what he thought some of his friends would say, which also made us laugh a lot.

We cracked ourselves up with memories we had made, like Dylan driving me nuts with his singing, which prompted him to grab a stick for a microphone, stand up like he was on stage, and start singing loudly, putting on a funny performance. We laughed so hard, and I smiled from ear to ear. Proud as punch I am of my baby and his ability just to be himself no matter the situation. I joined in singing with him, and for a moment, we forgot all about our surroundings, the exhaustion, starvation, dehydration, aches, and pains. I loved being in that moment with him.

When he finished the song, he sat back down, beaming and proud of his performance. I clapped loudly, smiling broadly. We sat hugging and smiling as reality seeped back in. I could see the fatigue wash over him, and I asked if he wanted to lay down.

He shook his head to say no.

"I'll come and lay down, too," I said, encouraging him to rest.

"Ok," he said quietly.

We stood up, and I put my arm around him as we walked to our bush bedroom. Sitting on the ground for a few minutes before we laid down, I looked at him and said, "I love you very much." He replied, "I love you, Mum," wrapping his little arms around me. I gave him a big hug and reassured him he was going to be okay and he would be rescued. I patted his back for a while, then we laid down, and within minutes, he was asleep, and I dozed off not long after him.

When I opened my eyes later, I could see the sky was still bright blue and clear. The sun was high, and at a guess, it was around 1 p.m. I turned my head to look at Dylan and saw his face had turned a light purple-blue colour. I let out an almighty scream.

Horrified, I leapt up onto my knees and bent over him. My heart felt like it had stopped. I screamed and screamed, thinking he had died, when I saw his body slightly jump. His eyes opened, and he looked at me and said, "Mum, are you okay?" looking scared. I grabbed and held him so tight I didn't want to let him go. I was so overwhelmed with fear and relief I couldn't even speak. "Mum, what happened? You ok? Why did you scream?"

Calming down, I took a minute to think about what to say. I didn't want him to know I thought he had died. "I thought you weren't going to wake up from your sleep," I replied awkwardly.

"Well, that means I would be dead, Mum," he said without hesitation.

My stomach lurched; my heart felt so heavy I thought it had gone into shock. I conceded to telling him the truth and said the only word that would come out, "Yes."

He looked at me, really looked at me, and finally said, "Mum, am I going to die?"

Oh, my heart, it just couldn't get any worse or feel any worse than in that moment. I wouldn't give up. I hugged him tightly and said a firm, strong "No." Regardless of the doubts I had, I wouldn't let us die.

I couldn't let go of him. I just wanted to keep holding him and never let him go. I was beside myself with fear. He pulled away, and I could see the light purple-blue colour still showing on his face. I felt sick. FUCK! Someone, come and rescue us! For fuck's sake! I knew Dylan was on the wrong side of survival now and wouldn't be able to last much longer. We needed to get out of here, NOW!

I had to get him water and get him up and moving to get the blood circulating. I told him, "Let's get up and get some water." His eyes lit up, and we stood. I held his hand and said for him to follow me. I honestly didn't know where to go, but I knew the water was down the mountain, not up, so that's where we had to go. We were blocked in several directions, so we headed down the gully and navigated the drops and hopefully would come across one of the creek beds. I just hoped like hell it wasn't one of the dried-up ones we had already found.

Down we went, sometimes walking, sometimes sliding down carefully and slowly on our butts. I kept talking to Dylan to ensure he was ok and with me, both in body and mind. We were nearing the bottom after an hour or so. The bushland again looked different, so I had no real way of knowing where we were, but I knew it wasn't where we had been before or near the shack where the dirty puddle was. In some ways, I was relieved we weren't back at the shack, as I knew there was nothing there— no source of water or help.

Dylan needed to stop a few times to rest. Each time I scanned his face to check his colour, my heart hurt a little more, seeing the purple-blue tinge. I felt so desperate I couldn't breathe. The realisation that my baby was close to dying was just too much for me to comprehend. I used the emotion to push hard and do what it took to keep him alive. I would do anything I could to ensure he survived.

After another hour of descending and zigzagging, we reached the bottom of the mountain we had been on. The sun had dropped from straight up in the sky to about halfway down to the ground, so I guessed it was around 3 p.m. I looked around me and tried to calculate our next move. Directly in front of us was a wall of thick shrub blocking our path. On both sides of that, the mountain was high and straight up, and then in the other direction, another section looked like it was lighter in shrubbery.

I took a few steps forward to look closer through the shrubs and quickly stepped back. There was a gigantic drop-off right in front of us. My heart raced, thinking how close we could have come to plummeting to our death. I turned quickly to Dylan, raised my hand, and said not to take another step forward and stay still. My mind raced; what now? How the hell do we get out of here?

I took a few steps back, tightly holding on to the shrubs on either side of me. I edged back further with my arm behind me, holding Dylan back. To my right, the bush was so thick with shrubs you couldn't even see through it. To my left, the shrubs in front of me were part of the mountain that, from the look of it, led to a flatter section. It was steep, but it didn't have the deep drop-off that was in front of us.

I was so scared and fearful that one of us would fall. I couldn't see what was beyond the flatter area. For all I knew, the mountain could have dropped off beyond that, too. Dylan was standing a few feet behind me, asking me what we were going to do and which way we were going.

I explained to him what I could see and said our only way forward from here was to slide on our butts through the edge of the shrubs to the flatter space. He didn't say a word. He just stood there and looked at me. He could sense I was scared, and I could feel he was too.

"Stay behind me. When I move, you move. If I stop, you stop, ok?" I said. "We need to take this real slow, remain calm and focused. It'll be okay. We'll make it safely; just take it easy."

He nodded wide-eyed, and I could see the worry all over his face.

I took a few steps, carefully one in front of the other. I was a few steps into the shrub when I turned around to check on Dylan. He was standing still. I said, "OK, take a few steps towards me and stop right behind me."

He moved slowly forward, and I said, "Good. Nice and slow. You're ok, we're ok."

When he got within a step of me, I turned back and stepped forward another step. I could see the bottom of the drop; we were dangerously close to it. I moved slowly away from it, trying to push through the shrubs and flatten them to make it easier for Dylan to follow me. I could feel the vines that had latched onto the shrubs tearing at my bare legs. I motioned for Dylan to take another few steps to stay with me.

My heart was pumping so hard it felt like it would bounce out of my mouth. I was so scared. I turned to Dylan and smiled, "Are you ok?"

He replied, "Yes," without confidence. I pointed to the left side of the flatter section of the mountain. I knew he would be able to see it more evident now,

"That's where we're heading," I said, with as much confidence as I could muster up. I don't think it worked. He knew I was petrified.

"Ok," he said, mirroring how I felt.

We stepped slowly forward, and I sat on a shrub for a minute to allow him to crouch down, rest a bit, and give me time to assess the ground and our position to the drop-off, which fell away in a rough zigzagged way. I said to Dylan we would keep moving slowly forward on our bums, but inch to the left also to navigate through the shrubs and clear the areas where the mountain free-fell into a drop.

As we continued the slow and calculated decline, our nerves were through the roof. I lost count of how many times we asked one another if we were ok. Every few metres, I'd turn to check on Dylan, making sure he was right behind me, and give us both the chance to breathe normally rather than holding onto our breath.

Remaining on our bums, I stopped to survey the next few metres. The side of the mountain had a slight uphill climb before we could reach a flat space on a small hill.

Every move was a physical challenge. Every direction was a gamble. There was no clear view from where we were to tell if we were heading to safety or even more danger. I stopped again, contemplating our surroundings. The high mountain walls and drop-offs were confronting. We were in a cul-de-sac. Ahead was the mountain wall covered in greenery, and the sides, one of which we were on, angled in various ways to the bottom, where I could see a large tree lying dead amongst the rocks and bushland.

I told Dylan to shuffle forward a little to sit behind me so he could peek over my shoulder to see what I was looking at. His eyes nearly popped out of his head when he saw the drop. "Now you can see why we've had to go really slow," I said.

"Yeah, Mum, I don't want to fall down there. What would happen if I did?" he asked.

Not wanting to say the obvious, I said, "I would find a way to get down to you. No doubt you would have a broken bone or two if you took a tumble down there," I said, playing it down but not sugarcoating it entirely.

"How would they rescue us if we did fall?" he asked.

I answered with a flat-out lie, "Someone would spot us walking the trail," knowing full well a fall would mean instant death.

To lighten the conversation, I added, "Imagine someone coming across us down there? They'd be like, WTF?!" We both laughed, and I focused back on the task, which was changing the subject.

"You need to move back behind me, baby. I need to move ahead and feel the ground to make sure it's stable so we don't slip, and check the shrubs are strong enough for you to hold on while we move around this ledge," I told him.

I started sliding on my butt to the left, grabbing the shrubs as I went, giving them a yank to make sure they held strong before Dylan took hold of them. The shrubs were dense, and I had to climb over them, around them, and push them apart or flatten them to make a passage for Dylan.

Lifting my leg over one to push it down, my foot became lodged in it. I had to wiggle my shoe around with a great amount of force to free it. It was like being clamped by the jaws of life. Dylan watched my every move closely to make sure he lifted his butt over the shrub or moved around it in the same way I had. Navigating his legs, he shifted his body to follow me successfully.

Coming to a section so dense we couldn't push through it, I said to Dylan, "Ok, you wait until I tell you to move." It was a dicey section extremely close to the ledge, and I had to be sure that by moving to the left further, there wasn't a sudden dropoff I couldn't see.

Feeling my way, I kept my hands on the ground and lifted my butt up and over the shrubs for a couple of metres, then looked back at Dylan and told him it was safe to follow, but to make sure he did exactly as I had and put his hands in the same spots as me. With each move I made, I pushed down with my hands, feet, or butt to test the ground, and also leave a partially flat area so it was easier for him to follow.

The process was repeated for another ten or so moves until we reached the part of the mountain where we had to climb up and onto the flatter area. It was steep and critical for me to figure out the best way to tackle it. I sat for a while, processing and mapping out how we could do it.

Dylan looked around me to see what I was looking at. There were no trees to hold onto in that section. It was mostly dirt and short green grass, so there wasn't anything we could gain some leverage with. We talked it through, each giving feedback to suggestions the other came up with.

In the end, I decided to climb up it first while Dylan moved to the spot I was sitting in to watch how I did it and follow me if it all worked out. I wasn't so much worried about the climb; it was more so leaving Dylan and being so away from him. I reiterated he had to stay put and not move a muscle until I told him to follow. He reassured me he would.

I stood up, leaning forward, facing the side of the mountain and the steep climb. I dug my hands into the Earth, grabbed onto dirt and grass above my shoulders, and brought my left foot up, digging it into the side of the hill. I pressed and pulled downwards, testing the soil to ensure it would hold firm. Hugging the mountain, I dug my hands in and pushed myself quickly up on my left foot to get as much height as possible, before moving my left and right hand up further, bringing up my right foot, and digging it in to take my weight.

I worked quickly and carefully, repeating the move until I reached the top, where I laid on my back for a few seconds, out of breath and on the ground, completely drained. I could hear Dylan ask if I was okay, and I barely was able to say, "Yes, I just need a minute." My heart was beating so hard it felt like my body was shaking, and my throat hurt to swallow. I forced myself to sit up, uneasy that Dylan was not beside me.

Dylan was looking directly at me. "Ok, you see how I scaled the hill? I didn't stop, and that's what you need to do. Exactly like I did, ok?" I panted.

"Yes, ok," he said.

"I'll lean down and grab your arm when you're within reach of me, ok?"

He nodded, determined. What he didn't have in energy or strength, he made up with willpower.

He moved as I had done over to the base of the hill underneath me. I said, "Ok, when you're ready, dig your hands in, kick the toe of your shoe into the hill and start climbing."

Dylan took a few deep breaths, psyching himself up and shaking the blood through his hands, feet, and legs to get them ready to move. After a few seconds, he made his move. I laid down and reached over as far as I could, outstretching my hand to him, ready for him to grab hold of me. He would need to climb up about halfway himself before he would be able to grab my arm, and I could hoist him up the rest of the way.

Encouraging him, I cheered him on, "Good boy! You're nearly to my hand. Keep going!"

Within a short time, he slapped his hand in mine, and I grabbed hold of him tight, pulling him to me as he used his other hand and feet to reach the top. Relief flooded through me. "Thank goodness," I breathed silently. We both laid flat on our backs, and in between pants, I said, "Good boy…You did it!" He never said a word. I could hear him breathing heavily, puffed out and trying to catch his breath.

We were now in another unknown obstacle course some-where in Mount Royal National Park, but it was a good feeling to know we were, for once, off the mountain we had been stuck on for the past eight days. Another checkpoint and milestone passed.

I knew we needed to keep moving, but neither of us had anything left to keep going with. We literally didn't move. "We might rest here for the night," I told Dylan.

I could barely hear his reply of, "OK," becaue he was so shattered.

The sun was low, and the area became shady as daylight slowly faded away. I estimated we'd have maybe 2 hours left of light if we were lucky.

"When you wake up, baby, we will walk down this hill and find some water." The hill had sapped all his energy. The most he could get out were single-word replies.

The ground here was covered in short grass under tall trees. It was way too tall for me to break any branches off to cover ourselves for the night. Honestly, we were so dehydrated and exhausted that we didn't care.

We lay beside each other, barely talking, until it grew dark, and we fell asleep. I had no idea how long we slept, but when I woke up, it was still dark, and the night air brought on the now-familiar shivering.

Dylan woke up what seemed to be seconds after me and said, "Oh, it's nighttime, Mum."

In the darkness, we could see each other and a small area around us, but the night cast a cloak over everything beyond. Through the trees, I could see a few scattered stars. I can't remember falling back to sleep, but the next thing I remember was again waking up. I rolled to face Dylan, who I could hear was sleeping soundly. I put my arm over his chest to make sure his breathing was consistent and was immediately reassured.

I thought of the discolouration on his face, and fear, guilt, shame, and sadness stabbed right through my heart. If I ever got out, I knew in all my days, I would never forget that moment. No matter how tired I felt in the morning, we would move on and keep searching for water.

Throughout the night, we woke and slept. At one point, I told Dylan it felt like we were sliding down the mountain. He agreed. I didn't understand how it wasn't slippery. We moved back a metre or so only to wake up feeling like we had slid down again. It was very confusing, and I was questioning if it was even happening at one point. I couldn't and still can't explain it, but it continued throughout the night until the morning.

Chapter 12

LIQUID GOLD

Day 9 TUESDAY

Opening my eyes, it was daylight—still early, around 6 a.m. I thought. I quickly looked at Dylan, who, thankfully, was still breathing and sleeping. Relieved, I said a silent prayer of thanks to whoever was out there listening.

With my arm around him, I thought of water, specifically about finding it. No fog was around, and the sun shone through the trees. I could see the sky was clear, and there were no clouds. There was no chance of rain, so there was no point waiting around for that or digging any holes to capture water.

Thinking back to the day we came across the dirty puddle, I wished we had drunk it, no matter how bad it tasted. It was better than no water at all. Apart from the urine I had drunk, in eight days, all we had taken in was the moisture from the leaves on our seventh day out here.

I slowly lifted my arm off Dylan and quietly sat up. Dylan opened his eyes a smidge and said, "Mum?"

I said, "It's ok. I'm just sitting up. You try and keep sleeping." Without another word, he turned his head and fell back to sleep.

I looked at the strands of grass around where we were lying to see if there was any condensation or moisture on them, but they were all bone dry. Disappointed and feeling let down, I knew there wouldn't be that option available to us.

Dylan stirred again and said, "Hi, Mum," casually.

"How are you?" I asked him.

"I'm so sick of being in here. I want to get out."

The knot around my heart tightened, and I said, "I know, baby. Me too. I want to get you out of here so bad."

He asked, "Will we ever get out of here?"

I moved between hopeful and doubtful a hundred times a day. We'd been here for far longer than I ever thought possible. It was completely surprising and incredibly disappointing that we hadn't been found or seen anything resembling a search party or even another person in the whole nine days.

We had climbed up and down the bloody mountain in all directions, and no matter which way we went, we never seemed to find a way out. The bush held us captive and hid us away from the rest of the world. There was nothing I wanted more than to get him out. Whenever I doubted we would be found, I worked hard to focus on the hope we would find a trail that would let us escape, or that our knights in shining rescue armour would see us and take us home.

I never shared these thoughts or doubts with Dylan. He needed not to lose hope, so that's what I focused on and told him. We were going to get out. Until then, I needed to find him water. We were both extremely weak now, and soon, climbing, walking, or even moving would become impossible. I didn't know how much longer we had, but I knew it was coming closer. Even breathing during the smallest amount of activity had become challenging.

We kept our spirits up by joking and laughing, creating stories, and dreaming up scenarios around our rescue. Deep down, I knew we wouldn't find the original track we came in on. Even though the bush sometimes looked the same—it deceived us. The mountains were everywhere, the trees and bush thick in every direction as far as your eye could see. We ended up somewhere else every time we climbed up, down, left, right. It was like being in a giant Venus Fly Trap plant. Regardless, we talked about and dreamed of finding a track and just walking straight out of this unpredictable and unknown land.

I stood up. My butt had become numb, and I needed to stretch my body and have a good look at where we were and which direction to head in. To our left was the big drop to the bottom, a big no on that one. To the right was where we came from, so there was a negative on that one also. Behind us was a giant mountain, and in front of us was a mystery. Our surroundings were covered in thick bush, engulfing everything, hiding everything.

Dylan stood up and asked, "Mum, what are you looking at?"

I said, "I am looking in all directions to see if I can spot a road or trail, a house or cabin, or any sign of life." There was nothing, just mountains, trees, and shrubs. Even though we were probably halfway between the top and the bottom of the park, we couldn't see anything different from when we had been higher up. "We're going to start heading further down so we have a better chance of seeing a trail, and so we can find water. Let's go."

We headed off, walking side by side. It was nice not to be on our hands and knees crawling on our butts for a change. It wasn't challenging terrain, only a very slight slope down. We didn't need to hold anything; we just had to watch our footing and stay together, but I noticed we were walking and moving at a far slower pace than we had been, and we hadn't gone far before we had to sit and rest.

Resuming walking, we came to another flattened area flanked by trees and shrubs. I could no longer see far into the distance above the mountains as before, so it was safe to assume we were now completely off the bottom and onto ground level.

Now, to work out which way to go. It didn't really matter, I guess; it all looked the same, so I said we would continue going straight ahead and see where it led us.

We had walked and trekked through the bush, around trees, and over fallen logs until I spotted a creek bed between two trees. I pointed in the direction to show Dylan and silently thought, "PLEASE let there be water, please!" We quickened our pace a little in hope. The nearer we got, the further away our hopes drew. The creek was quite deep, scattered with bush debris, and dry as toast.

Standing at its edge, I scanned up, down, and everywhere. There was not one bit of water, not a drop—nothing but rocks with green moss. I was beyond devastated. I told Dylan we would follow it in the hope of it leading to water or leading us out of the park. The sides of the creek were relatively high, so we walked along the edge for what felt like forever to find a lower section where we could climb safely down without jumping and risking an ankle or leg on the rocks.

We continued to walk along the bank as it snaked through the bush. It was the same height the whole way along for as far as you could see. Up ahead, it bent to the left. What was beyond that was anyone's guess.

Both sides of the dried and rock-lined creek were surrounded by thick bush. Some sections of the bush were right up against the side wall and threatened to topple down the bank and fall to a rocky death.

Walking on, we noticed a few big rocks lying close to the side of the creek wall we could climb onto and lower ourselves safely into the creek bed. "We can use those rocks for an easier climb in," I said, thinking out loud.

Arriving at the section, I told Dylan that once I was in, I would reach up to him and help him down. I sat on the creek bank and swung my legs over the edge like I was sitting on the side of a swimming pool. Dangling my legs, I reached the nearest rock with my foot. The large grey rock was covered in moss, so I was careful not to slip while I lowered my legs onto it.

Once I got safely on the rock, I laid down upon it, hugging it, slipped down the side and landed on the creek bed floor. I assisted Dylan as he followed onto the rock and slid down safely into my arms. Standing in the creek, it smelled damp, mouldy, and rotten. Green moss was everywhere, suffocating everything in view—the walls, rocks, fallen trees, everything. The rocks were all different sizes and shapes. None would have been small or light enough to pick up, only to step on or over.

We followed the creek bed in what would have been downstream had it been full of water. We continued for hours in search of water with the sun directly above us, filtering through the giant trees. It was quite cold in the creek, so we welcomed the sections that were not as thickly covered by the tall tree branches so the sun could penetrate freely and offer some warmth onto the ground. Sometimes, we stopped in these places for a few minutes, letting our bodies heat up before continuing.

We kept busy moving to stave off the cold and were met with a section that divided into two directions. Taking my mum's advice that I could hear in my head, we went left. I'm not sure why Mum had or believed in her motto, but I hoped her philosophy would bring us some luck, and, in a sense, it was my way of carrying her with me.

Travelling left, the creek bed looked the same as it had for the past few hours. Some sections of the wall were far higher than they had been, though, and in these areas, I could barely see the surrounding bushland. Some sections had no boulders or large rocks to climb over, which was nice for a change, not that it would have mattered; we would have climbed over them.

This creek bed bent and snaked left, right, and left again the whole way along. Still, there was no water, just slippery, gooey green moss wherever you stepped and looked.

As we rounded yet another bend, I stopped and couldn't believe what I was looking at. The creek ended abruptly at the base of a mountain wall. It was too slippery and too high to climb out of. The side walls were ridiculously high, and no rocks lay beside them, so we had no hope of climbing out. I felt like sitting down and crying. I was so disappointed. All that time and wasted energy for nothing. We now would have to go back to the intersection and head right.

Dylan was looking at me, waiting to see what we would do. "Do we need to go back?" he asked.

I looked at his drawn, exhausted face; all I wanted to do was say no, but there was no other way. We had no choice. "Yes, we do. Not all the way back, just to the part where there were two directions."

His shoulders slumped, and he put his head down and sighed deeply. He was feeling as disappointed and disheartened as I was.

We were exhausted, starving, dehydrated, and running on air alone. Yet somehow, we had to find something more in the tank to keep moving. "Can we rest for a bit before we start walking back?" he asked.

"Of course, we can." I felt desperate for a rest, too.

Looking around to find the least-covered mossy rock I could see, I told Dylan to sit and rest for a bit. I remained standing, worried I might not get up if I sat. I stood near Dylan, and he leaned his weight against me. I put my arms around his shoulders and said, "I know you are thirsty and hungry, but I'm very proud of you and how you're keeping going." He didn't reply. He was completely spent trying to catch his breath.

I thought about calling out as usual, but my throat was bone dry, and even the few words I'd spoken to Dylan were challenging. With such dense bushland around us, there was probably no point anyway.

After a while, I patted Dylan on the back and said, "When you're ready, we'll start the journey back." Not that I physically wanted to move, and I'm sure he didn't want to either, but in a quiet voice I could barely hear, he said, "Ok, let's go," and lifted his head and gingerly stood up.

I told him when we got back to the intersection point, we would stop and rest again before setting off in the other direction. I wanted to hurry up, reach the intersection as quickly as possible, and start heading in the other direction for two reasons. One I knew there was no water in this direction, and Dylan's situation was becoming direr and direr by the second; two, I wanted to see if there was a possibility it could lead us out.

Up and over rocks, twisting around fallen trees, we headed back with minimal conversation. Finally, after a while, we reached the intersection, and Dylan sat to rest while I stood again.

Looking in the direction we needed to head, I could tell it was slightly uphill and lined with a lot more moss, debris, and larger boulders. "Fuck, this is going to take some effort," I thought silently, but with a good deal of wishful thinking to motivate us both, plus some big encouragement from me, we would push through and find some water, hopefully.

Dylan was understandably far from keen to move when it was time to get up and start walking again. "What if that way ends too?" he asked me.

"Well, then, we need to climb back somewhere and find another way." He didn't say a word, just nodded.

Following the creek uphill, we held onto the sides of boulders, rocks, and tree branches to assist us in getting through or over whatever obstacle was in our path. Eventually, we reached the top, puffed and exhausted. It felt like it took an hour or more to reach this point, and I could see from looking at the sky that the day was getting on, and we had maybe four hours of daylight left. We'd been climbing, trekking, and getting through obstacles all day and had yet to find any water, using more and more of our energy.

Reaching the top of the hill, I was relieved to know it didn't come to a dead end like the other direction had. Instead, we were able to keep following it, so far at least. The creek went downhill and bent around to the left. Beyond that, I was unable to see what lay ahead.

Dylan didn't stop; he said he wanted to keep going. While he was happy to keep moving, we did. I realised, like me, he may have been finding it challenging to rest and get started again.

We had to climb over branches; we couldn't go under them because of the boulders underneath. One boulder was so big it blocked the whole creek bed. We had to navigate and climb up on other smaller boulders to get on top of it, then scale down the other side.

The first of the smaller boulders would have been as big as a picnic table. It was a strenuous process of me climbing up and then turning around to help pull Dylan up over each boulder. As we neared the top of the most enormous boulder of all, I could see it was covered in moss, and I thought to myself, "Shit, it's going to be very slippery."

If I went up the boulder first and turned around, I could reach out with both arms to grab Dylan's hands. Locking hands, it was hard work keeping my grip on the boulder while wiggling back and pulling Dylan up, but he did a brilliant job of putting his feet on the right areas to push up towards me.

We sat perched on top of the slippery, giant rock, checking out the path ahead, which was cluttered with debris. There were so many fallen trees and branches that they sat in layers on top of one another. Some branches still attached to trees leaned in precarious angles by their roots. The only way down this giant boulder was by sliding down it on the other side. I wasn't confident I was going to be able to reach up to Dylan once at the bottom, so I thought it best that he face the rock and, while holding my hands, slide slowly down with me hanging onto him for as far down as possible, until he could get his foot on one of the smaller boulders a fair way down towards the base of it.

Getting safely down the remainder of the way, he reached the bottom and stood back. I could see his shirt was wet and green with moss and hoped it would dry before nightfall so he wouldn't be too cold. Hopefully, his body heat from walking would dry it out enough.

Now, it was my turn. I shifted my butt forward to move more onto the slope of the boulder and slid steadily down it. I could feel the wet, cold moss as Dylan directed me down, encouraging me. "Mum, you're not too far. Keep moving. Go slow." Hearing his voice guide me and reassure me I was nearly at the bottom was sweet.

Hearing Dylan's voice getting closer, I lifted my head to see how far away the ground was, stretching my legs out to feel for something to put my foot on, when suddenly I felt my body fall straight onto a boulder.

Landing directly onto my back, I laid there for a few seconds, not moving. I could hear Dylan asking in panic, "Mum, are you ok?"

My first thought was the sheer terror that if I had damaged or hurt myself enough, I would not be able to get up and keep walking. I lay there with this overwhelming fear, feeling a pain in my back, and too scared to move. Thinking about Dylan, I tried lifting my head. "This was not the place where he'd become stuck or die because I could not move," I thought determinedly.

I opened my eyes to see him directly in front of me, eyes wide and scared. I must get up, I willed myself. Taking a deep breath, I pushed my arms down beside me and sat up. The pain shot through my back, but thankfully, it wasn't agonising. I was beyond relieved I could move and hadn't done some big damage.

I stood up on the slippery, wet moss. No section was free from it. It was everywhere. My shoes must have become slippery and wet from it when I came down on the boulder. "I'm ok," I said, leaning on the boulder for support.

I stepped cautiously, moving away from the mossy rock, wondering how there could be so much wet moss but no water. The creek was dry, and there wasn't as much moss on this side of the boulder as on the previous side.

"Let's keep going," I said, directing us through the tree branches and rocks, avoiding the ones that looked unsafe or were heavily covered in moss so we didn't hurt ourselves further. Sometimes, it meant we could walk carefully over or around the rocks; sometimes, we had to lean down and hold on to the rocks with our hands; sometimes, we had to squeeze between them, turn on our sides, and manoeuvre through them like a crab.

My eyes constantly scanned, searching desperately for water as we walked for endless hours. Dylan asked for a break here or there, and I tried to encourage him to go a little further so we could cover as much ground as physically possible. I was determined to find water. We had to. At this stage, it was the only thing keeping us going.

I didn't know how much longer Dylan's 9-year-old body could sustain exerting even the smallest amount of energy loss. He was dehydrated, hungry, and exhausted—I couldn't expect him to continue for too much longer.

I had to keep moving, and for me to keep moving, I unfortunately needed to keep Dylan moving, too.

Distracting him, we talked about the birds and the calls we heard. We couldn't see them and didn't have the energy or care to look for them. Our heads were down, focusing on where we were stepping. Where it was a tricky section to walk through, I'd share my strategy with him to keep him mentally engaged. He didn't talk much; when he did, it was to ask if I thought we'd get out or to convey his hopes of finding water. I silently sent a plea to some higher power that we would.

The day was passing fast; it was already about 3 p.m. Only having a few more hours before the daylight faded, I began to think about where we could rest for the night. We couldn't walk much further; we were exhausted, my back was sore from my fall, and we were dragging our bodies and feet.

Looking around us, I noticed that the edge walls had become knee-high low, and the rocks were more spaced out. The creek bed had become wider than it had been previously, and it was much flatter as well. I was curious about what this meant. Walking side by side now, I shared my observation with Dylan, who looked around, agreeing.

With the creek walls being so low, we could see the ground of the park and bushland again, which we hadn't seen all day. The thick bush once again surrounded us instead of the high creek bed walls. From the bushland, the big mountainous terrain sprouted.

We continued following the creek bed. The light was fading fast, and the sun could barely penetrate through the trees that had become thicker. We hadn't felt cold for most of the day, continuously moving. Our shirts had dried from our body heat, but my shorts around my butt were still damp. -Dylan's shorts and T-shirt were so dirty that they were hardly recognisable. I looked down at my clothes, and they hadn't faired the eight days we'd been out here well, either.

I noticed to the right of us that the creek wall had veered out wider in a half-circle shape with a flat section in front of it. The creek had come to an end. Disheartened, I couldn't believe it led us to nothing more than the same landscape we'd been trapped in the whole time. I felt like crying and wondered, what now?

To the left side of us was a thick bush surrounded by mountains which we had no energy to climb. Regardless, there wouldn't be water up higher, so there was no point in even considering it. To the right side of us was the same, a small flat part that once would have been a swimming hole, now empty, that led uphill to another mountain. Directly in front of us, we could step out of the creek and onto the bank by holding on to a fallen branch and levering ourselves up and out.

Dylan sat on the ground, resting. I stood looking around, trying to figure out what to do–and where to go. I remained standing, knowing there was a high possibility I would not get back up again. I had never felt so deflated before. In every direction was a mountain, thick with bush and trees.

The only hope was a small hill I could see through a sparser spattering of trees. I wanted to leave Dylan to rest and climb it myself to see what was beyond the mountain, but the risk of being separated was too great. I reached out for his hand and said, "I know you're tired, but we need to keep going to find you water." He reached out his hands for me to grab and pull him up.

"We are going to walk towards the top of the hill up there," I said, pointing in the direction we were heading. Even before reaching the top, we could already see it was leading us nowhere. It was more of the same, with no tracks or water to be seen.

We wandered aimlessly downhill, zigzagging with no plan, no direction other than down in the hope of finding water.

No section was flat. Every few steps we took, we had to change direction to go left or right around a tree, bunch of shrubs, or other obstacles, until it became too steep to walk, and we had to move down on our butts again.

Dylan remained behind me most of the time unless I was holding a branch or shrub so he could get through. I regularly turned around to check on him and continued to keep him motivated with conversation. I'm sure the only word he really acknowledged and processed was 'water'.

At times, pushing through the scrub was difficult. When we finally reached the bottom of the hill itself, not much had changed; all we could see was this wall of uneven brown bushland.

We stopped for a moment, looking around to spot any way forward. Dylan was still standing behind me because there was not enough room for us to stand side by side; the bush was that thick.

Continuing, we came to a section that on either side of us had a slight drop into bushland, flanking a thick patch of green plants with red leaves. I had never seen them before, and they caught my eye, mainly because of their colour. Up until then, everything was pretty much brown. These plants had brown trunks with several shades of lush green leaves.

I stepped closer to them, and my shoe sunk a little into the soft ground. To my surprise, I could see moisture on the red leaves as I looked closer at them. Excited, I snapped one of the branches off and turned to Dylan, amazed. "Look! There is water on these!" I looked up and around, thinking that the sun could not reach this area due to the density of the thick bush in this part of the park.

"I will lick it first to check for any effects or reaction before you try," I said, licking the leaf and feeling that tiny drop of moisture on my lips. Oh, it was heaven! Dylan was staring at me, anticipating either a reaction or the okay for him to have a go. There was no foul taste, so I told him to sit down and start licking as many leaves as possible. Forever the kind, caring, and sweet boy he is, he said, "Mum, you lick them too!"

We sat amongst the patch of plants, picked off the leaves, and drew in as much moisture as the plants offered us. Each branch had four large leaves the size of a cos lettuce leaf. Finishing in one area, we'd shuffle along to the next. It was like we were harvesting, leaving the plants bare of their leaves behind us.

The moisture wasn't enough to quench our thirst, but it was enough for our lips and tongues to feel wet, to offer some relief for our dry mouths and throats, and to provide some hope and joy.

After taking all the moisture available, we kept moving around the bush, which was too thick to see through or beyond it. The ground had become even softer, and as we walked, we noticed the dry dirt ground transforming into thick grass. It wasn't like the regular grass in our backyard; it was more of a light green shrub grass. The thin strands tickled our legs as we walked, and our shoes began sucking to the earth beneath the grass, to the point that with every step, we had to pull our legs up with a bit of force to free our shoe.

Behind me, Dylan had his shoe stuck and gave his leg a hefty yank to free it, which sent the shoe hurtling off over my head. At first, I had no idea what had just flown past me, and I turned to Dylan and asked, "What was that?!" He told me it was his shoe and said, "I'm so sick of this. I am sick of being in the bush. I'm sick of walking." I stopped and put my arms around him, giving him a tight hug. I knew we were both at breaking point.

After ten more steps, I looked to the left of me and noticed some trees and branches that had fallen onto each other across what looked like another creek. I pointed it out to Dylan and said we should check it out. "Let's find your shoe first, though, hey?" No matter how hard we looked through the grass and shrubs, we could not locate his shoe.

Dylan was over it and said, "It's ok, Mum, I don't need it." I was concerned about what he could step on without a shoe to protect his foot, so I searched for a bit longer without any luck.

Walking towards the fallen tree branches, we climbed through, over, and under them to get to the creek bed we had spotted that was snaking around to the right. It was much the same as the others we had followed previously, with different-sized grey rocks, some clamped together, others spread out, fallen tree branches, and moss on some rocks and the side walls. Again, no water. Where it went, who knew? All I hoped for was water and that it wouldn't end and would lead us somewhere different—hopefully out!

Dylan walked in one sock and one shoe. It didn't seem to bother him. Honestly, he was too dehydrated, exhausted, tired, and starving to care.

I kept asking him if he wanted to wear my shoe, and he kept replying, "Mum, I'm fine," but I could tell by the tone of his voice when he responded he was frustrated, and understandably so.

"Ok, as long as you know if your feet hurt and you want to wear my shoe, you can," I said. We followed the creek and reached a section where the left side turned into thick bush, and the right side went downhill and continued to another bend.

The downhill section was clear, with no trees, no rocks, and no fallen branches, and the ground was covered in red and brown coloured leaves, like a bed of autumn leaves. It was pretty to look at. We sat to rest, and with our lips and mouths more moisturised, we spoke more and commented on what a cool camping spot it would be.

"Daniel would love this," Dylan said.

I agreed; he would love it. "He would have a fire right there next to the creek and pitch his tent right here on this spot." We smiled, thinking, visualising Daniel being there with us.

I checked Dylan's foot and asked him if he wanted to put one of my shoes on, but he was adamant he didn't want to. Fifteen minutes later, after a decent rest, I said we should keep going. Dylan reluctantly agreed.

Standing up, I looked at Dylan's foot and asked him again if he was sure he didn't need my shoe. He shook his head, and so I laughed, "You look so funny with one sock on and one shoe!" He looked down at his feet and laughed, too. Putting one arm around his shoulders, I said, "Come on, let's go," smiling.

We started trudging the few steps down to the bottom of the hill and into the dried riverbed. I looked left to where it bent around and stopped dead in my tracks. I could not believe my eyes! "OMG!!! I found some!!!" I yelled.

A few steps behind me, Dylan said, "What did you find?"

I turned with bright eyes and the biggest smile on my face. "Water! I found water! You can have a drink! I finally found you some water!" I said, elated.

Dylan was beside me in a second and looked at the little puddle I was pointing at. He was shocked. It had been that long since we'd seen water; it was like finding liquid gold. The feeling of seeing it was indescribable.

It was only about the size of a big bowl but filled with clear, beautiful water. It was so clear it was like a fish tank; you could see the tiny white rocks in the bottom of it.

We looked at each other, smiling, and quickly hugged each other. I couldn't stop saying, "OMG, we found water!"

We bent down, kneeling at the puddle. It was too small for Dylan to put his head in it, and I wouldn't have wanted to risk any spilling over anyway. Every drop was priceless, and I wanted him to drink every single bit.

It was so gorgeous to look at, crystal clear, not even a leaf in it. For a few moments, I thanked whatever higher power had made sure this puddle remained here and drinkable just for my little boy.

I didn't want to dirty the water, so I asked Dylan, "Now, how will you drink it? I know, let's pick a leaf off a tree to use as a scoop." I figured there was more chance of getting every bit of the clean water without all the dirt from his face and hands. I picked a fresh leaf off a nearby tree and handed it to him.

Dylan said, "Mum, I can't believe we found this."

I replied, "I know! I am so glad we came this way." We sat down on the rocks next to the puddle. I was still in shock we were looking at water! Rolling the leaf into a scoop, I told Dylan to put his head over the water, and I'd scoop up the water for him.

Excitedly getting into position, he lowered his head and opened his mouth. I delicately and slowly scooped up the liquid gold and brought the leaf to his lips, pouring about 2 ml of cool water into his mouth. Scoop after scoop, I could see in his eyes the relief filling his withered little face. He was in pure heaven, enjoying every drop. It was the best feeling. I was thanking my lucky stars we had come this way.

After a few more scoops, he lifted his head, looked at me, and said, "Mum, you have some."

I felt guilty about drinking any of the water. I wanted Dylan to drink it all, so he had more chance of survival. His kindness and consideration lit up my heart. He still thought about his mum even in sheer desperation for the water.

I considered it briefly, and realised he had a point. I should probably drink some too, because my survival was essential to his survival. I bent over the puddle, which by this time was around half-full, and took my first drink of water in nine days. To be able to swallow actual water and not just moisture was better than anything I could describe. The cool trickle down my dry throat was nothing short of bliss. I had a few more scoops before continuing to scoop more into Dylan's mouth. He was beaming. "This is good!"

I began to think. If we were able to go nine days without water, now that we had found some, surely we could survive another week or more while continuing to search for more and a way out.

I don't think so, Michelle.

"Have more. I want you to have as much as possible," I told Dylan. I continued to scoop until about a quarter of the water left. He sat back content, and I said he could have some more later after a rest.

Looking around us, the sun was still shining brightly in a clear blue sky. I couldn't see the sun directly due to all the thick bush, only parts of the sky here and there. I thought it could be around 3 p.m.

A black and orange butterfly started fluttering near us, and I put my hand out to see if it would land on me. It didn't, but to see something so pretty, beautiful, and peaceful only enhanced the special moment.

We sat and rested on the slope near the water, and I told Dylan he could drink the rest of his water whenever he was ready. "Mum, and you too," he said, smiling.

I wrapped my arms around him, telling him how much I loved him. With his little arms wrapped around me, he looked up at me and said, "I love you too, Mum." What a moment! I could feel the emotion well up from inside me.

Our bellies still grumbled nonstop for food, but we had long ago learnt to ignore them. Sometimes, we would even laugh at the sounds. They were so loud at times it was like thunder crashing around us. For a lot of the time, it was the only sound we heard apart from a few birds or a tree branch falling occasionally.

We both laid back, looking up at the small section of the sky we could see, and the next thing I knew, I was waking up. I looked at Dylan beside me, who was fast asleep. It had become my routine when waking to quickly scan his chest for movement to confirm he was still breathing. For an instant, my heart felt like it had stopped from the sheer fear I wouldn't see signs of life. Thankfully, he was ok.

I had no idea how long I had slept, but the sun still seemed like it was shining brightly, so I must have guessed the time to be later than it was earlier on. The area we were in was mostly covered in shade, giving the illusion it had been later.

Dylan stirred and opened his eyes. "Are you ok?" I asked.

"Yes," he said, sitting up and looking around.

"I think we have a few more hours of daylight left," I told him. "Let's go drink the rest of the water and continue along the creek to see where it takes us."

I didn't need to ask him twice; he got straight to his feet, and we walked over to the puddle. It was still a sight to behold, sitting there clear, ready, and waiting for us to drink.

I picked another leaf off the same tree as before, and we sat down. I continued scooping the water into Dylan's mouth. After taking a few scoops, he leaned back and told me it was my turn. I took a few scoops and gave him another five before having another few scoops. We continued like that until the last few scoops, which I made sure went to Dylan.

Taking water in gave me renewed hope and determination to survive this ordeal. Finishing the water was a bit disheartening, but it encouraged us to find more, especially since we knew it wouldn't be long before we became dehydrated again with dry mouths and cracked lips from further trekking.

Even though we weren't overly excited to walk for hours again, the water had at least given us more energy to proceed.

The creek was much the same as previous ones we had followed—filled with debris, littered with hundreds of obstacles, and twisting and turning. Sometimes we were able to walk side by side; other times, I led Dylan behind me, cautiously calculating the path. At times, it was like crawling and contorting our bodies through branches like the laser beams in the movie Entrapment with Sean Connery and Catherine Zeta-Jones.

We had walked and followed the creek for hours, not speaking much, just focusing on safely walking. My throat and mouth had quickly become too dry to speak much, and I was sure Dylan probably felt the same. I only spoke if I had to or to check on him.

It had been an enormous day of walking for us, and we had reached our capacity and limit. To the right, I noticed a flat section of the creek bed like a step-up platform. I pointed it out to Dylan and said we would rest there for the night. Even though we were both desperate to get out of here and not have to spend another night, we had to stop. Our bodies were done for the day.

I looked on the ground for any visible rocks and sticks that I needed to remove so we didn't have them stabbing into us throughout the night. Even with the water we had drunk, neither of us needed a bathroom area anymore, so I only needed to prepare a place to lie. While I worked, Dylan asked again, "Mum, are we ever going to get out?"

I stopped and reached out my arms to him. Wrapping him up in them, my heart sunk even further, but I said positively, "Yes. Tomorrow, I'll work out which way to head, and we'll get out of here." I honestly don't know if he believed me, but in that moment, I said whatever it took for him to believe he would make it out and be ok. Every time I heard the desperation and worry in his voice, it broke my heart into a million pieces.

As I was hugging Dylan, I felt the pain seep through my back, and I winced. I released my arms from around him. I told him, "I'm going to turn around and lift my shirt. Can you check for any marks or anything on my back?"

I didn't have the energy to stand, so I just shifted uncomfortably to turn around on my bottom. I lifted my shirt and heard him gasp. "OMG, Mum!" he exclaimed.

"What? What is it?" I asked, worried.

He said, "Your whole back is black and purple!"

I quickly pulled down my shirt and thought to myself, "What the fuck were you thinking, Michelle?!" upon hearing his concern.

"Does it hurt?" he asked.

"A little, but I'm totally fine, so don't worry," I told him.

"It's your whole back, Mum. Do you want me to massage it for you?" he asked.

"No, I'll be fine. I can still move and walk easily, so it's ok. You rest," I answered, smiling at his sweetness. He had always been the kid who offered help to anyone first. I was so proud of him. Over the years, I've lost count of how many strangers, teachers, family members, and friends have said to me how polite and well-mannered he is. He'd even helped some new neighbours move their things into their unit and always held open lift doors. He is a fantastic kid.

The daylight had nearly faded, and we began to hear the sounds of the night—crickets, birds, and, at times, tree branches cracking and the dull thud of them falling to the ground. We could also hear movement in the bushes, but it was far away, so we weren't panicked by it. The air was cold; thankfully, however, there was no wind. It was completely still.

Underneath us was mostly dirt and very short grass. Some small clumps of a grass plant the size of your hand and moss were here and there. Even though our surroundings were quite moist, there were no puddles, wet leaves, or anything to draw moisture from.

The area was more spaced out here, with more gaps and flat sections of the creek bed. The depth and width of the creek bed here were also different. I had no idea what that meant, but I hoped it would be easier to walk tomorrow and hopefully lead to the way out. "Wishful thinking," I thought wearily.

The sky offered no moonlight, and we were engulfed in pitch black sooner than anticipated. I told Dylan to lie down, and I would huddle against his back to keep him warm.

He laid on his right side and got as comfortable as possible, and then I laid down and draped my arm around him. "Goodnight. I love you," I said.

"Goodnight, Mum. I love you too," came his soft and sleepy reply. Within a few minutes, we both were asleep.

We slept on and off, waking when we were uncomfortable or cold, only to remember there was nothing we could do about it. Knowing there was nothing we could do, we'd fall back to sleep, only to hear a rustling or noise close by that forced me out of sleep and into instant alert with my heart racing. With nothing to see but the blackness, fatigued, I would doze back off to sleep.

Of all the nights we'd been out there, this was the one and only night I dreamed. I was in a game, laying on the ground on a big flat piece of wood that was lifting me up into the air by a hydraulic pole it attached to. Once in the air, it would spin and turn. I knew I needed to stay on it for as long as I could. It kept taking me up higher and higher, spinning a few times to try and dislodge me off it, then dropping to the ground.

I clung to it, holding on for dear life, knowing that if I fell, I would be splattered on the ground. At the highest point, the platform lifted me, and I felt my whole body jolt and jump as if I were about to fall in the dream. My eyes sprung open, awake to the blackness of the night. I couldn't move. It was like I was too scared to move in case I fell off the ground where I laid, and I said aloud, "I am going to fall."

I felt Dylan move next to me. "Mum, are you ok?" he enquired.

I was aware of my surroundings but still felt the remnants of the dream. It was like I was simultaneously walking the line between sleep, being in the dream and consciousness in the present. I could still hear myself saying I was going to fall.

"It's ok. I had a weird dream," I told Dylan.

My daughter, Sarah, used to say to me, "Mum, you have weird fucked up dreams," she was right too. Happy with my response, Dylan got comfortable again and drifted off to sleep.

I lay thinking about the dream. It affected me somehow, but I couldn't figure out why. For once, I found some comfort in the sounds of the bush around me, even the bird that sounded like a baby crying—it initially unnerved me, but now we just laughed about it and said, "There goes that baby crying again!" I could hear another bird far away; it had a long, drawn-out whistle sound.

When Dylan was awake, we'd try to guess what the different birds looked like and come up with crazy images that made us laugh. Humour has always been a big part of our life. We believed we were naturally funny. Being out here, it had kept us sane. It was my last thought before dozing back off to sleep again.

Chapter 13

THE HELICOPTER

Day 10 WEDNESDAY

When I woke up, it was just on daylight. The air was cold, and there was fog in the distance. It was low and thick.

I was relieved I did not have another dream; the last one still haunted me.

I lay there trying not to think of the dream with my arm over Dylan until he began to stir. The cold throughout the night, or maybe me yelling out, had unsettled him. He'd had fitful sleep.

He opened his eyes, turned to face me, and said, "I don't want to be in here anymore."

I could tell he wanted to cry. I wanted to cry, too. The exhaustion, hunger, dehydration, and feeling of being uncomfortable every day for now, ten days straight had depleted our minds and bodies. I cuddled him, trying not to let my emotions brim over. "We will find a way out," I said, determined.

A big part of me thought we would have to get ourselves out. I had lost belief we'd be rescued. We hadn't seen or heard a helicopter or any sign of rescue for days. Even if we had heard one, my throat was so dry it was sore, and I had stopped calling out days ago.

The bush was so big, the scale of it enormous. We were two tiny people in an impossible and remote situation. We needed to keep going, somehow. My hope was to find more water, and my goal was to find a way out.

Dylan sat up to work out the discomfort in his body. Despite my best efforts to remove debris, it was impossible to remove all of it. The ground was covered in sticks, rocks, leaves, and shrubs. I removed what I could each night, but it was still lumpy, bumpy, hard, and debris dug into our skin, leaving indents and impressions.

I stood up, brushing off the dirt and debris, and stretched. I could feel my pants and T-shirt were damp and cold against my skin. I assumed it was due to the ground being moist in this area. My back was still sore from the fall I'd taken and not helped by sleeping on the ground. As I moved my body around, I was thankful there was no real pain, and to be honest, in our circumstances, it was the least of my problems.

I looked at Dylan stretching out. His red T-shirt was rainbowed with an assortment of brown from the dirt and green from the moss. He looked like he was wearing a tie-dyed shirt. "My clothes are wet," he said.

"Yes. Mine are, too. Not to worry, they will soon dry," I replied.

I looked through the trees to spot the sun and figure out what time it was. The sun shone through the trees onto different sections of the ground directly where we were standing. The area and ground around us were shaded, the trees and bush surrounding us so thick with trees and ridiculously tall, so finding the sun through them was a challenge. I didn't want to waste time trying to get ourselves out of here today. With no need to go to the bathroom and no options for food, there was nothing to do but begin walking as soon as Dylan was more awake. We walked side by side, mostly without talking. The rocks remained small, so walking over and between them was easier. The only real obstacles we had to navigate were the fallen branches.

Dylan lopped along with one sock and one shoe and maintained he was fine each time I asked him if he wanted one of my joggers. As it was a reasonably easy walk, sometimes I'd lead, sometimes he led. Whenever we came to a tricky section, he'd wait until I worked out the best way to navigate it.

The creek bed became quite broad, and I found myself pleading with a higher power to please not let it come to a dead end. Throughout the hours we walked, we were covered in shade. I still hadn't spotted the sun; all I could see from time to time was a light stream of sunlight penetrating a branch.

Beyond the thick bush on either side of us, we caught glimpses of the hills and the mountains that encapsulated us. We walked for what felt like forever, constantly scanning for another puddle, before the creek began levelling out, looking like it ended and became bushland again. "Please, no! Not again," I begged inside of my head.

The closer we got, the more we could see that the creek became a track that led into the bush. It was the first time we had spotted anything even resembling a track since we had become lost. I was cautiously hopeful and dared to wish that, by some miracle, it might lead us back to our original walking trail, the campground, and our car.

We stepped out of the creek onto the flat part of the ground and focused on the dirt trail I spotted, which snaked further ahead. I hadn't told Dylan until now in case it was a false alarm. I didn't want to get his hopes up for them to crash down again.

The trail was clear, overgrown in parts, and littered with bush and branches, but it was a path! We both became excited as we followed its windy lead. All I could think about was where it would lead us. "Please lead us out, please lead us out," I kept repeating silently.

Snaking around bends, the path led us to the beginning of another dried creek bed. Then, to my surprise, a pretty little black and orange butterfly began fluttering around us again. "It's back," we both said in unison, watching it dance around us. It looked exactly the same as the other one we saw when we found the puddle of water, the same colours, the same size, everything was the same. I hoped that was a good sign that we were headed in the right direction. As quick as it came, it left to fly up and over the trees and out of sight. "Maybe he is lost too," I smiled, and Dylan giggled.

Out of all the creek beds we had encountered, this one was obviously different. It still had rocks and fallen tree branches within it, but the difference was that there was not one grey rock with moss on it; they were all pure grey. It was a nice change not to walk so cautiously, trying not to slip.

The creek bed headed upwards for a while, and then we approached a bend. The creek walls had begun to form and were around knee-high, and the first signs of moss started showing on the rocks again.

Up ahead to the left, we spotted a strange rock. As I got closer, I couldn't believe what I was seeing! It was a bath FULL of water! I stood without moving in total shock. It didn't seem possible. The amount of water in it was like we were looking at a mirage. It was huge! Like a regular-sized bathtub and filled ¾ of the way up with water.

I looked at Dylan to see if he could see it or if I was dreaming. His eyes were lit up, and he was smiling. "Can you believe this?" I asked him. It felt like we'd struck it rich.

We walked over to it, and I knelt, bending over it. "Can we drink it?" Dylan asked eagerly.

"I think so," I said, gazing at the water, which was crystal clear, just like it had been in the puddle. The only thing I could see in the bath besides the water were tiny tadpoles.

"What are they?" he asked.

I explained what they were and assured him they wouldn't hurt us.

I wasn't about to let a few tadpoles get in our way and dunked my whole head in the water, drinking as much water as I could before coming up for air. Dylan followed my lead and only pulled his head up to breathe, smiling from ear to ear.

"Keep going; drink as much as you can! There is more than enough for us both," I said excitedly. I didn't want to stop, and it looked like Dylan didn't want to either. We kept dunking our heads and drinking until we felt sick.

I took a breather and looked to the right of the bath and noticed a flat piece of land covered in pretty red and brown leaves. It took me a while to realise I was staring at another track.

I stood up and pointed at it. Dylan, who had lifted his head to breathe, saw me and got to his feet. "Is that a track? Where does it go?" he asked, buzzing with anticipation.

"I'm not sure, but we will follow it and find out. Not yet, though; we will keep drinking this beautiful water first," I said.

I wanted to empty it. No matter how thirsty we were, I knew that was impossible unless we stayed here for a few days, but we would do our best to drink until we couldn't take in one more drop.

We continued to drink, taking some breaks in between until our bellies had so much water in them that they were bloated. Feeling our mouth, tongue, throat, and lips so moist was sheer delight.

I had stopped drinking and was leaning back when I felt something biting at my arm. I flung my arm out, and that's when I saw a leech had somehow attached itself to me and was endeavouring to suck my skin. Shuddering, I flicked the slimy bloodsucker off with one of my nails and quickly scanned my body and Dylan's to make sure there were no others. My body shivered. I knew they weren't poisonous, but all the same, the thought of them sucking our blood wasn't inviting.

I was reluctant to leave the bath at first. I didn't know when we would come across water again, and given the bushland's history of deceiving, it made me fearful we wouldn't find our way back to the bath once we began walking again. But the lure of a track and the hope of either walking out of there or seeing another person to help direct us was too much to pass up.

Now that I had taken in some water, I hoped my throat wouldn't be sore for long and I could start calling out again soon. I told Dylan how the track reminded me of the one we had originally got lost on. He thought the same and said, "Imagine if it was, Mum." Oh, how I hoped that to be true. Just imagine.

"Ok, well, let's take another few gulps of water and fill ourselves up to the very brim before we begin walking again," I said.

After another few minutes, we couldn't drink anymore. Dylan was leaning back, full and completely satisfied, which made me overjoyed.

"What a relief it was to find this liquid treasure," I thought. I will never forget the moment I thought he had died, waking up to him pale and purple and blue. Just remembering it, I second-guessed myself. "Do I take the chance to get out and walk away from this water source or stay?" I questioned myself.

Before I could bounce it back and forth in my mind, Dylan got to his feet and asked, "What if the track leads us nowhere?"

"Hopefully, it doesn't, but if it does, we can walk back here for a drink and then follow the creek bed uphill to see where it goes. We need to keep moving if we want to get out of here," I explained.

"Ok," he said, and we began walking toward the track.

I knew being on a track was our best chance of being seen or rescued. I kept my eyes and ears open for any possibility. The water had restored our confidence, and with one last and longing look at the water, I thought, "I hope I am making the right decision."

The track was wide, and we walked along together. Apart from some smaller branches, it was relatively free of bush debris. As it wound around to the right, I could see the sun peeking through the trees and glimpses of what looked like a beautiful blue sky.

Walking along the track under the shade of the tree canopy, our hair around our faces began drying out again after being soaked in the water bath. The track went from wide to thin, then widened again. The ground changed as well; there was more dirt showing through the short grass. The trees were tall, and it was like we were walking through a forest rather than the thick bushland we had been in.

Visibility ahead became more apparent, and we could see the mountains and hills fully exposed to sunlight. "This is different," I thought. Then, suddenly, both Dylan and I stopped in our tracks and looked at one another. Was that a helicopter?

"I can hear that!" The sound was different from the one we had heard several days ago, but there was no mistaking it; it was definitely a helicopter, and it sounded like it was coming closer and closer.

We scanned the sky wildly, hoping to see it and, more importantly, it could see us! The trees above were still shielding us. We spun around, searching for it. It sounded so close it felt like it was right above us somewhere. The rotors were deafening. "Where is it?!" I yelled above the noise. Then, it came into view briefly. I could see it was yellow and red. "OMG! It's a rescue helicopter!" I yelled to Dylan.

We waved our arms around and screamed, "We are here! We are here!" I grabbed hold of a tree trunk to try and shake it so the rescuers could see someone below them, but they were too large and didn't budge. We jumped up and down, waving madly and yelling out to them.

The helicopter quickly turned around and flew away from us in a split second. "No! No! No! Please don't leave us! We are down here!" I yelled as moist tears welled up in my eyes.

Even though I couldn't see it, I could hear the helicopter heading away from our direction. Tears began running freely down my face as I turned and hugged Dylan hard, "They are looking for us! But they can't see us through these big trees."

Dylan asked, "Will it come back?"

I hoped it would with every part of me. "Yes!" I said confidently. My hopes were up, and my belief had returned.

"We need to keep moving and try to find an open space so when it does come back, we have a better chance of being seen," I told him. With renewed energy, we walked quickly, following the trail uphill, downhill, left, and right, through thick bush, and sections that looked like a forest out of a scary movie. We didn't deviate from the trail and stuck to it like glue.

We could still hear the sound of the helicopter, although it was faint and now seemed a long way from us. I imagined they were scanning another area for us. As we walked, it didn't come closer again, just the opposite; it grew very faint, and then there was silence.

"What does that mean? Aren't they looking for us any- more?" he asked with worry lining his face. "Don't worry, it is searching somewhere else. Remember, this place is huge, and there's a lot of ground to cover. It will be back. Just keep listening out for it. In the meantime, let's hurry and get to an area where they can see us," I urged him.

We had been walking for hours at this point, and with all the excitement and physical exertion, we had depleted our energy levels. I needed to stop, and Dylan did too, but he was reluctant to in case the helicopter came back, and we were still under the canopy of the trees. "We will hear it, don't worry, even if it's far away. We'll only stop for a few minutes."

Feeling more assured, he said, "OK."

"I won't miss any chances for them to rescue you."

We sat in the middle of the track, not budging from it. I smiled and put my arm around his shoulder, hugging him before lying on the ground to rest.

I opened my eyes. Realising I had fallen asleep, I sat bolt upright and quickly looked for Dylan. He was sitting in the sun, drawing in the dirt with his finger and humming to himself. I scrambled to my feet and stood up. "Was I asleep long?" I asked.

"No," he said, but I knew he didn't have any concept of time being so little.

I checked the sky; the sun was directly above us. "You need to get out of the sun and move into the shade more," I told him, worried about how long he'd been out in it. He stood up, and we continued to walk along the track to under the shade. Fifty or so steps along, the track veered to the right, and I could see a big clearing ahead of us. It was littered with the same grey river rocks we had crossed over on our first day on the original track.

As we came to the end of the dirt and grass track, we met another dried creek bed about the width of a typical residential street full of rocks the size of watermelons. For the first time in ten days, we were in a clearing big enough to be seen by anyone flying above us!

Aside from the rocks, a few boulders roughly half a car's length in diameter littered the area. Up the hill a bit, the rock area forked off to the left and right. I felt the familiar dilemma of wondering which way we should walk. In both directions, the rock corridors were lined with thick bush.

My first instinct was to remember what Mum had always taught me and go left. I took another few steps onto the rocks to get closer to the middle of the section and see if I could see anything in either direction before we made a choice. To my utter amazement, I spotted another bath. "OMG! We've found another one, Dylan!" I exclaimed.

Dylan walked straight to it, eager to have another drink. I followed him and saw that it was much the same as the other bath, not as clear, but you could still see what was in the water, and this time, joining the tadpoles were tiny silver fish!

I remembered Daniel, my eldest son, was around six months old when we went to New Zealand with my parents and two younger sisters. My mum had cooked up these tiny little silver fish and added them to her dish. I wondered if they were the same. They looked the same. "We could eat them if I could catch them," I thought.

The bath was a little bit bigger than the previous one. "Can we stop here for a while? Maybe the helicopter will come back and see us?" Dylan asked.

"Yes, we can. Let's drink some water!" I said happily.

We dunked our heads in and gulped, breathed, and gulped some more until we could drink no more water. While examining our direction choices, Dylan asked, "Which way will we be going?"

I told him we should take a bit of a walk further up the hill before deciding.

Stepping over the large rocks was tricky; we struggled a little, trying to keep our balance on them. We had to be careful not to get our feet lodged between them or slip. Making our way up further, we looked down toward the left side and could see that the rocks met thick bush, so we eliminated that route straight away.

We had to walk over the rocks past the bath some way to see down the right-hand side. A few metres past the tub, we could see the low riverbed walls leading to another track. Dylan spotted it too and said, "Mum, look, it's another track," pointing at it. I was so relieved. I had felt discouraged initially seeing yet another dried creek bed, but this one proved to be a real gift with water *and* a track!

I asked Dylan if he wanted to drink more water before we made our way to the next track, and he didn't, so we began walking. With our ears tuned to the sky, hoping to hear another helicopter, we clambered over the rocks. I was hesitant to leave the clearing, but at the same time, I thought it could be days before we heard another helicopter. My gut feeling told me we should try the track.

Off the rocky creek bed, we stepped up onto the track. It was flat and covered in different coloured leaves—green, brown, and red. In some sections, the sides of the track were so thick with bush I couldn't see past them, and in other areas, it was so sparse I could see the high mountains off in the distance.

Dylan said, "Imagine if this one takes us back to the track where our car is."

I chuckled and said, "Yes, imagine that!" As I walked, I could feel my mobile phone and the car keys still safely in my pocket. Oh, how amazing would it be if we could go home! The starvation we both felt wasn't as bad as the dehydration we had experienced, but it was still bloody awful.

"If this track does get us back to the car, what will we do?" Dylan asked.

"Well, I think we will get in the car, start it up, plug in the phone to charge it, and drive until we get reception and then call everyone we know to tell them we're OK," I replied. We had laughed and talked numerous times about everyone's reactions to us making it out of here. It was a story that kept our hopes up. What a story it was, too. Even I couldn't believe it, and I had lived it!

We walked silently, both lost in our thoughts about home, family, friends, and life outside Mount Royal. The track bulged wide and narrow and wide again, leading us this way and that way, up and down.

The sun had disappeared, and through the sparser areas, I could see grey clouds had begun to appear. I hadn't given much thought to the weather during the day. All my attention had been on finding water. Once we had done that, all my thoughts were of getting out. I'm not sure how long we walked, but it was probably hours when the track ended, meeting with another creek bed.

There were the same grey rocks, but no water bath I could see, unfortunately. Again, we had to choose left or right. Both veered around a bit of a bend, so we would have to walk along if we wanted to see what lay ahead before choosing. I dreaded making the decision—either could be a mistake if there was no clear indication of which one looked like the best option.

Stepping down onto the rocks in the creek bed, I looked along the opposite side of the bank to the left. It was more of the same thick, overgrown bushland. Scanning towards the right, I noticed a small gap between the bushes and a flatter section. I pointed it out to Dylan and said, "I think that's another track," as I headed towards it.

Walking closer, Dylan said excitedly, "Yes, it is!"

Relief washed over me. Thank goodness.

Making our way over to the other side of the creek bed, we noticed the track wasn't as wide as the one we'd just come along. The bush hugged this one far more, and if we hadn't had the water we had fortunately come across, I have no doubt we would have been too mentally and physically exhausted to spot it.

We could see the right side of the bush bordered the track, but the left side was more open. One section in particular opened into a large, flat, open area scattered with hundreds of fallen leaves.

The temperature turned cool as the night air brushed up against us. I told Dylan this area would make a good camping spot for the night. The grey clouds were still visible, and I guessed it wouldn't be too long before it was dark again.

"We will sleep on the trail just in case in the morning we get confused or unsure of which way we have come, and also in case someone comes walking along. Let's sleep with our feet pointing toward the dried creek," I suggested.

We sat side by side in the middle of the path, buggered. We were both happy to stop and rest. Sitting with our feet facing the creek, I said to Dylan, "I think you need to put my shoe on."

"No," he said.

"It will keep you warmer; it's going to be quite cold tonight, I reckon," I explained to him.

That made sense to him, so he agreed. I pulled it off my foot, and he took it and put it on, commenting on how warm it was. "See, I already warmed you up!" We laughed and laid down, looking at the daylight beginning to fade away.

"Do you think we'll see a helicopter tomorrow?" Dylan asked.

"I hope so," I replied.

I woke up in the dark and felt raindrops on my face. There were no stars and no moon. More and more drops landed on my face. Dylan woke seconds later and asked, "Is it raining?"

Feeling a few more constant drops, I replied, "Yeah, it is starting to rain, but I don't think we will get much out of it. It rarely rains in here. Try to go back to sleep."

He nestled into me, and I closed my eyes but soon felt the raindrops getting faster and heavier. "Oh shit, baby. We need to move."

We stood up, and I grabbed Dylan's hand so he stayed close. I tried to look around and remember our surroundings. I knew to the right side of the track was thick bush; the left side was flat with spaced-out tall trees.

Holding Dylan's hand firmly, I guided him to one of the large trees on the more open side, hoping it would shelter us. It began to pour down. There was no escaping it. We were soaking wet within seconds. In the pitch black of the night, we couldn't easily see anything to hide under. We slowly moved about, trying to find somewhere suitable to stand and block out the rain, but it was useless.

I told Dylan to stand against a tree trunk so it blocked the rain from coming in at him from one side, and I stood on his other side to protect him as best as possible. It was pelting down now, harder and heavier. I could feel water cascading over my face, running through my hair, flooding my clothes and socks, and filling up my one shoe. The dirt and grass quickly became soggy and squishy under my foot.

The rain was relentless, and the tree wasn't offering any kind of buffer from it. Dylan was shaking against me. We needed to find somewhere else. With one hand again firmly holding

Dylan's, I led with him behind me and stretched my other hand out to feel in front of us as I walked.

I felt my leg hit something, and I fell forward. I let go of Dylan's hand to break my fall on a tree log lying on the ground. I could feel pain shoot up my shin. I was lucky I hadn't fallen face first; it could have been much worse had I not reacted so quickly.

I stood back up, letting Dylan know I was okay, and took his hand again. "Be careful of the tree branches on the ground!" I shouted, above the sound of the rain.

Moving slower, I tried again to search for some shelter, but the darkness and the water running into my eyes made it virtually impossible. The rain was not easing, and my one shoe had gotten stuck in the mud and sucked off my foot. I said, "I've lost my shoe." I didn't even bother to stop and find it; there was no point; it was too dark. I could barely make out the shadow of a bunch of trees in front of us with a bit of a ledge like a garden bed. We stepped onto it, and I got Dylan to wiggle forward to sit between me and the garden. I wrapped my legs and arms around him to try and keep him warm.

We were shivering uncontrollably like we hadn't ever shivered before. To distract us, I told Dylan, "Imagine how much water we will have to drink in the morning from all this rain. We can walk to the river rocks and drink some of it before we start walking again."

"OK. We should try and find your shoe too, Mum," he said.

After a while, the rain began easing and turned into a light shower. I told Dylan to rest his head on my knees and try to get some sleep while he could. Within a few minutes, he was asleep.

I could still feel his body shivering; my body was shivering too. He woke up often; it was the most uncomfortable I had felt yet, so I knew sleep would evade me, too. All we could do was wait it out.

Eventually, the rain stopped. It was quiet, apart from drops of water falling from the branches to the ground. I looked out into the darkness and thought of all the fresh rainwater we could drink in the morning from the puddles and rocks in the creek banks. There would be no shortage of water.

I rested my head forward, leaning onto Dylan's neck, and closed my eyes. I began to think about how far we had moved from the track and stressed about being unable to find it again in the morning.

Chapter 14

MY WORD

Day 11 THURSDAY

I opened my eyes to see and was thankful it was light. It was still very early, around 5:30 a.m. to 6:00 a.m. I guessed. I lifted my head, my neck stiff from the position I was in and from the cold. Dylan was still sleeping.

The rain had passed entirely, and we were sitting in a blanket of fog that covered the ground entirely and blocked the view of the distance. It was like another world. Now and then, a cold drip of water splashed onto the back of my bare neck or arms. I was drenched, still cold and shivering, but nothing like I had been the night before.

Waiting for Dylan to wake up, I sat in the same position with him nestled between my legs until my back started aching, and I had the now rare sensation of wanting to urinate. After all the water we had drunk, I wasn't surprised. I tried to ignore it and block it out for as long as possible as I didn't want to wake Dylan, but the feeling escalated, and I needed to move.

I slowly moved my legs, trying not to disturb him, but it instantly woke him up. "Sorry. I tried not to wake you, but I need to stand up."

He lifted his head and looked around, saying, "The rain has stopped."

"Yes, it rained hard for a while longer after you dropped off to sleep, and then it finally eased up," I said, standing and stretching. It felt so good after being crunched and huddled up all night.

I was busting to urinate. "I'm going to walk behind this tree and go to the toilet, OK?" I asked Dylan.

"OK, Mum," he replied.

As I walked around the tree, I noticed we weren't sitting in a garden bed at all; it was a small hill surrounding a few trees that were bunched together. How different things look in the daylight, I mused.

Relieved I was finally able to empty my bladder, I pulled my soggy underwear and pants up. My socks were like ice and made my feet feel cold and numb. I mentioned it to Dylan upon returning.

"Let's see if we can find your shoe," he said, walking to the area he thought it had become dislodged in. Nothing looked familiar to me. It felt like we had only gone a few feet on either side of the track, but I could have been wrong. Scanning the ground, I noticed a large log. "I bet that's the log I fell over last night," I said, remembering how relieved I was not falling face down on it. I could have woken up with a broken nose or chipped teeth!

"I found it! Here is your shoe, Mum," Dylan said proudly, holding the wet, muddy jogger.

"It's wet," he said, looking at it sadly. I laughed.

"Thank you. I'm just glad you found it; being wet doesn't matter. We are wet anyway!" He smiled and handed it to me to put on.

"The path is there, too," he announced, turning and pointing. I craned my neck to see, and sure enough, there it was. What a relief. "Great spotting!"

"Let's walk down to the rocks and see how much rainwater there is for us to drink." I had prepared myself to see many full puddles from the amount of rain the night before, but when we got to the rocks, there was no water to be seen. I was dumbfounded. How could that be?

Walking along, we spotted just three tiny puddles, each with maybe a cup or so of water in them. "Wow, I thought we would have found lots and lots of puddles! I'll pick a leaf to scoop up the water."

Within minutes, we had drained all three. "Maybe the water just doesn't stay in these rocks like I thought, but at least we had some more water to drink," I said, and Dylan agreed.

"Let's go back to the track. The fog has disappeared, so we can begin walking again. We'll take off our shoes and squeeze as much water out of our socks as we can. We won't be able to do much with the water in the shoe, but it will hopefully keep our feet a bit drier," I explained.

Being now one shoe short, I struggled to walk on the rocks and the track. It was covered in tiny little stones that hurt the bottom of my feet. I couldn't walk at the pace I was able to before giving Dylan my shoe. I asked Dylan, "How did you walk on this with no shoe?" He laughed and said it didn't bother him or hurt his feet. I was surprised; it was killing mine.

On the track, we took off our shoes and forced as much of the water out of our socks as possible before pulling the shoes back on our feet. Dylan handed me my other shoe and said, "You wear it, Mum. I can tell your feet are hurting when you walk."

"No, no, you put it on your foot," I insisted.

"No, you wear it. My foot doesn't hurt when I walk without it," he said, holding the shoe out closer to me.

I felt guilty but relieved to take it. "OK, thank you. Let me know if your foot hurts, and you can take it back. We can take it in turns wearing it."

I reached out and hugged him. "I love you so much," I told him. "How did I get so lucky to have the kindest kid of all?" I thought proudly as we set off.

Looking at the sky, we could see it was thick with grey clouds. No sun shined through as we trekked along. The track up ahead bent around to the left and wound up a steeper hill section. "OK, we will get to the start of the hill and stop for a minute to catch our breath, then keep walking straight to the top."

Dylan looked at the hill and said, "It's so high."

"I know, but it's part of the track, so we need to keep following it," I told him.

The bush on either side of the track had become thicker, and the path was littered with freshly fallen leaves. Some had a few drops of water on them, and I stopped to pick up every one I found along the way and tipped the contents into Dylan's mouth to try to keep his throat and lips moist, knowing it wouldn't be long until he became dehydrated again.

A few minutes later, we began the journey up the hill. No matter how far we walked up towards the top, we didn't feel like we were getting any closer to it. It seemed to go on forever. We needed to stop multiple times; the climb was exhausting, and we breathed heavily.

Eventually, we reached the top, tired and puffed out but relieved to have finally made it up. I stood trying to catch my breath when I heard the faintest sound of the rescue helicopter. We both became excited and hugged each other before looking in the direction of its sound. It was a fair way off, which gave us time to think through whether we should keep going to find a clearing or race back down the hill to where we had been, which was open to a degree.

I was torn. Walking back down would take a while as it was steep, and it wouldn't even guarantee being seen as there were still quite a few trees down there. Plus, all our hard work hiking up would be undone. Walking ahead meant no certainty we would find an open area again. We decided to push forward, stick to the track, and hope it led to a clearing large enough to be seen.

We walked as quickly as we could. The whole time, I second-guessed my decision to push forward. I could still hear the helicopter whirring in the distance, but it seemed like it was moving further away, which made me more confident to keep going rather than turn back.

The track wound all over the place, sharply bending left and right. Sometimes, it was flat, and sometimes we headed uphill further. The dirt, bare in some places, became short grass in others. We didn't deviate from the track, hugging it wherever it led us. I could tell we had climbed quite high up. The bush had thinned out, and we could see the other mountains far in the distance.

If not for the fact that we were following the trail, we would not have walked up this far. From my previous experience, I didn't have a lot of faith in us being rescued so high. We were still finding leaves here and there with a few drops of water in them, and I always stopped to make the most of the moisture, carefully picking them up and tipping them into Dylan's mouth.

As we came to a left-hand bend, the track and the grass disappeared, and we walked on bare dirt. We spotted some deep wheel indents in the ground. We both looked at each other, puzzled. They were far too wide and big for motorbike tracks. They looked more like monster truck indents. "Maybe they're from a large 4WD or tractor?" I wondered aloud. We walked in the indents along the track that was so deep they had large dirt shoulders. Dylan took the right side, and I stepped into the left one, and we walked a fair way before stopping to rest again.

Laying on our sides in the tyre tracks, I looked up and said through my heavy breathing, "Look. It has come back," pointing up to where the pretty little black and orange butterfly gracefully danced in the air us. Again, I wondered if it was the same one we kept seeing and if it had some meaning I was missing. It was truly beautiful.

We watched it delicately fly around us before it flew off into the trees and out of sight. We didn't have the strength or energy to speak, so we just looked at each other and smiled at the wonderful and welcoming experience.

Now and then, we could hear the helicopter draw closer, then further away. It felt like we were drawing away from it in the hours we had walked.

Going uphill was hard work, and Dylan commented on how tired he felt and didn't want to keep walking anymore. I understood entirely, but stopping wouldn't lead us out, so I suggested we rest for a while longer, but we would need to continue.

The trees were far apart from each other here and set back from the track a bit. After a few more minutes, I checked on Dylan and asked him if he wanted my other shoe. He didn't reply and laid still, too exhausted to speak. "Are you ok?" I asked. He nodded and slowly sat up. I could see he was struggling, and the look on his face told me he wouldn't be able to keep going for much longer. I stood and took his hand to help him up, and we walked hand in hand, side by side, not speaking.

Focused on the leaves hoping to find any remaining drops of water, Dylan stopped suddenly and pulled at my arm. "What is that?" he asked, pointing ahead of us. I froze, my heart raced, and the panic rose inside of me. I pulled him back quickly, saying, "Holy fuck!" There was a big, red-bellied black snake curled up in front of us with its neck and head resting on its body.

I am petrified of snakes. I remember my eldest son once brought home a snake. I came home from work one afternoon, sat at the dining table chair, and was asking how his day was and explaining mine when I noticed he had a very guilty look on his face. "Why do you look sus?" I had asked him.

With a cheeky grin, his eyes went directly to the floor beside me.

I felt my body go instantly into flight mode and jumped up so quickly from my chair that I knocked it over. I screamed and cried, running down the hallway away from it and yelling, "Get that out of here!" I was so upset.

He giggled and said, "It's fine, Mum, it's only a carpet snake."

"Why would you bring it home knowing how scared of them I am? Get it out."

It took ages for me to pluck up the courage to creep slowly back down the hallway to check that he had removed it. He held it like he was holding a cute little puppy! The hairs on my neck stood up, and I had to beg him to take it outside. He said a friend gave it to him. "I don't care; go give it back. You need to get rid of it," I had said firmly.

"OK," he said, still laughing hard at my reaction. Evil child, I had thought, smiling afterwards.

Walking slowly backwards away from the snake before us, I didn't know what to do. We couldn't go back now; we had to go around it somehow. My eyes darted to either side of the track. There was hardly any room to walk between it and the bush that bordered the track in this section. I needed to calm myself down to think, but I felt scared out of my mind.

Dylan could see my turmoil, but he was more curious than scared like I was. "Is it sleeping?" he asked. I thought it could be since it wasn't moving. Maybe it had come out of the grass to lay in the sun and get warm after last night's downpour.

I looked down at Dylan's shoeless foot and thought, thank god we hadn't stepped on it. The chances of it biting him were high. Red-bellied blacks were notoriously aggressive, but that was the extent of what I knew about them.

"Can it see or hear us?" Dylan asked.

I had no idea; I only knew I wanted to run.

We backed further away from it as quietly as we could. "I'm going to get a big stick," I told Dylan.

Looking at the bush behind us, I retrieved a big and solid stick. "Don't go into the grass; there may be more of them," I told him. I leaned down to take my shoe off and told Dylan to put it on to protect his foot.

"No, Mum, I will just stay on the other side of you," he said.

"OK, I will hold the stick down to the ground in case the snake moves towards us. If it does, I want you to run back the way we've come from as fast as you can," I instructed him.

It made no sense that having the stick gave me the courage to get past the snake, but it did. I took a deep breath and felt as ready as I would ever be to move forward and get around it. I quickly scanned the area to see if there were any more ahead and went through my plan to Dylan.

"Walk as close as you can to the edge of the track, don't go into the bush, and keep hold of my hand unless I tell you to run." I held the stick down to the ground, keeping it between us and the snake. "We are going to walk as fast and as quietly as we can past it," I said, feeling my heart thumping hard in my chest.

Keeping my eyes on the snake, we moved closer. With each step, it felt like my heart would lunge out of my mouth. I could feel the fear growing more and more the closer we got to it. When we were directly in front of the snake, my fear was so intense that I felt my heart would stop. I couldn't breathe. We walked fast without disturbing the ground or making a sound past the snake and went some distance beyond it before I stopped, letting go of Dylan's hand and letting out a massive sigh of relief. My heart belted against my chest as I took my eyes off the sleeping, undisturbed snake and looked at Dylan.

Even though we had gone well past the snake, I felt on edge and continued to scan the path and edges for more. I had what my mum would call the heebie-jeebies, so I held onto the stick to be cautious for a long time before finally releasing it and letting it fall to the ground. It was enough of a challenge to carry myself, let alone a long stick. Eventually, I relaxed as we put more distance between us and the snake, and my thoughts returned to the track.

The bush on either side of the track had begun falling further back, and we soon came to a fence line that ran along the right-hand side of the track. I could see it ran ahead of us for a fair way until it disappeared into bushland. Dylan asked what it was for. I didn't know, but it looked like it had been there for many years. We couldn't leave the trail to follow it in case we lost our way back. Also, the bushland looked thick, and I didn't want us to come across more snakes, which I thought would surely be lying hidden from sight. The track was safer, offered a clearer view of where we were walking, and had less chance of getting lost.

"The irony," I thought and half smiled.

We kept walking like a couple of zombies without knowing where we were headed. We kept moving forward in the hope that the Mount Royal National Park would release us. There was no emotion, no speaking, no nothing. My legs were on autopilot, one unenthused step after another, walking on, following the track.

We hadn't heard the helicopter for some time, and the excitement and joy I had felt from seeing it the day before and the hopes of being rescued felt like a long distant memory. Dylan didn't ask about it anymore, either.

We continued to walk and walk until coming to a complete stop. I couldn't process why we had stopped for a few seconds. I was in a daze. The track had ended, and we were standing on the edge of a road. An actual road!

"Mum, we're on a road," Dylan said, looking both ways up and down the road.

The awareness of where we were, how stunned I felt, and the dryness of my throat left me without a voice to even utter a single word in that moment. Then it hit me. We were out! The elation of being free of the bush came at me all at once. I began jumping up and down like a lunatic alongside Dylan. We were hugging and crying, and through cracked voices, yelled, "We made it! We made it out!"

We were amazed! We didn't know what to do first. It had been eleven days since the park had swallowed us up, and here we were after hundreds of hours of trekking, surviving freezing nights, starvation, and hell-raising moments, finally and impossibly on a road! We had gotten out—alive!

I couldn't explain, even if I tried, the feeling of knowing my baby was going to be *for sure* safe. It was incomparable to anything other than bringing my children into the world. The emotion was all-encompassing. I promised him I would get him out, and I finally did. I thanked the stars, moon, sky, and every higher being that I knew of, smiling and hugging Dylan.

Still hugging and ecstatic, we looked impatiently in both directions, wondering which way to go. Checking our location, I said to Dylan, "This looks like the road we took that led us into the Mount Royal."

He said, "Oh yeah, it is," and explained how he remembered seeing the fence across the road from the car on the way in. I remembered our conversation, too! We had talked about how it would be to live out here surrounded by so much bush and the animals you could have on so much land.

Thinking back, I recalled we saw a house just up the road. We walked along the road and came to a gate with a sign that said, 'Private Property. Do Not Enter.' I thought, "That's not going to stop me." I pushed the gate slightly and called out as loud as my throat would let me to see if anyone could hear me.

I waited a few seconds, and there was no response. I opened the gate fully, fearful that dogs might come bolting out and rip into us. "I will walk in front of you, and if I tell you to run, turn around and run as fast as you can and get behind the gate again, OK?" I told Dylan.

"What about you?" he asked.

"Don't worry about me. I will protect you and fight off anything that comes our way so you have time to get to safety, and I'll be right behind you," I said, hoping it wouldn't come to that.

We walked down the dirt driveway surrounded by bush, and my first thought was, "Oh god! We are back in the bush again!" I walked up the driveway, calling "Hello!" to no response.

A cabin and a caravan came into view, and honestly, it looked a bit scary at first. I thought to myself, "Oh please, don't let us survive all that we have to now walk into a killer's property." I kept calling out but couldn't see any movement. I felt confident there were at least no killer dogs roaming around since there was no barking. We also noticed there were no vehicles in sight on the property.

We walked toward the cabin, and I told Dylan, "Geez, it's going to be hard to explain who we are and that we have been lost in the bush across their property for ten nights!" I walked up the stairs and knocked and knocked. There was no answer or sound or movement whatsoever. I tried the door, but it was locked. I looked at the windows to see if any were open and told Dylan that if there was a way in, I could get him some food, and then we could walk back to where our car was and drive home.

I knocked on the door a few more times, tried the door again, and checked the windows. It was locked up like Fort Knox. We walked around to the back of the cabin. The windows back there were too high to reach, so we were out of luck there too.

I noticed a black recycling tub full of empty water bottles and a tap off to the side, attached to the cabin. I gave it a try, turning it. Yes, water! I collected two bottles from the tub, rinsed them out, and filled both before handing one to Dylan. We skulled both, and I refilled them.

"No one is here, so let's go back to the road and follow it to where I parked the car," I suggested. I knew it would be a bit of a walk, but it was a longer walk back into town, so it would be quicker to head to the car instead.

Walking side by side, I joked about the owners coming home and seeing us trotting down their driveway. We laughed and thought they'd be asking themselves, "Who the hell are these crazy loons?!" I could only begin to imagine what we both looked like by that point.

We reached the gate, opened it, walked out of the property, and shut it securely behind us. We crossed over to the other side and had only taken about ten steps when a white car came whizzing towards us. I dropped the water bottles and waved frantically at it. As it got closer, I realised it wasn't an ordinary car but a police car! Talk about the right place at the right time! What are the chances of that?!

As the police car slowed down, I ran through what I would say to them. Pulling up beside us, the officer closest to us on the passenger side said, "We have been looking for you both." Everything I was going to say didn't come out. I was speechless. I didn't expect them to know who we were. It took me by surprise. The same police officer who spoke got out of the car and asked if we were okay while opening the back passenger door for us to get in. Dylan jumped in first and slid along to make room for me.

All of a sudden, my legs buckled underneath me, and the officer grabbed me and carefully lowered me into the car. Somehow, my legs knew they didn't need to do anything more and just stopped working altogether.

I leaned over and hugged Dylan tight and began to cry. "You're going to be OK. We made it out," I sobbed hard. I could hear one of the officers asking us if we were alright and reminding us that we were safe.

"How did you get lost?" one of them asked. I was too emotionally overwhelmed to string a sentence together. I barely got out a few words.

"Don't worry, you're safe. Where were you headed?" he reassured me. I told him to where we parked our car.

He smiled and asked, "What were you going to do when you got there?"

I looked at them both and said casually, "Drive home."

They smiled and said, "Your car was towed away."

I was confused. "Why?" I asked.

"Well, the detectives had to smash a window so they could try to figure out what had happened to you."

I couldn't make sense of that.

"We're going to take you straight to the search and rescue set up not far from here to get you checked out."

I had no words trying to process that we were out and safe. Unbeknown to me, it was all about to get very real.

We arrived at the campsite where I had parked the car, and Dylan pointed, saying, "Mum, look at the campsite." It was crawling with SES volunteers, police, and an ambulance vehicle was also parked there, too. There were tarps set up, and I could see food and water under them. The police officers got out of the car and turned to open the door for us to get out as well. Suddenly, we were engulfed by a sea of people, all talking simultaneously and asking a thousand questions. It was hectic.

I don't remember much of what was said, but I will never forget two comments stuck in my head. One from a person who said that we only had a 20% chance of survival, and another from someone who said that the search for us was only going to go on for another two days before they would have had to call it off, presuming we were dead. I honestly didn't know what to think about it all as I looked into the dozens of curious faces around us. I was speechless at the fuss as the ambulance took over us from the police officers.

The paramedic asked, "Do you know what day of the week it is?" as other people asked, "How did you do it?" and "How did you get lost?" and a flood of other questions. Information, comments, and requests for details bombarded us. It was too much to take in, and the noise was deafening. I could barely speak more than two words at a time and held tight onto Dylan's hand.

The SES workers asked if we were hungry, and both of us nodded fervently, which produced a lot of comments about starvation and how hungry we must be. A worker handed us each a Subway roll and Dylan mowed into it until it was gone.

I took a big bite and nearly vomited. I couldn't understand it. I would have eaten anything after eleven days without food. I especially loved salad and couldn't understand why it tasted off. I tried to eat but couldn't stomach it. I just wanted more and more water and gulped it down before feeling my legs grow weak and give way again. I panicked, unable to hold my body up, so the paramedics carried me to the step at the back of the ambulance.

Emotion started spilling out of me, and no matter how much I tried, I couldn't stop the tears from pouring out and down my dirty face. The paramedic rested her hand on my leg, giving me comfort, and kept saying how incredibly brave we were and how amazed everyone was that we were alive and had walked out on our own. Hearing that, my thoughts went straight to the moment I discovered Dylan blue and purple. It was too much to express. The relief he was going to be okay was immeasurable.

Dylan sat behind me, and I reached out to him and softly patted his arm. My body had had enough and was weak and felt lifeless. My legs had this constant tingly feeling running through them. I could barely speak and couldn't move, but Dylan, to my amazement, was excellent. Physically, it was like he'd had nothing happen, the paramedic reassured me. He sat up in the ambulance, drinking juice and water, eating food he was given, and laughing.

After a thorough check, the paramedics explained to us, the SES, and the police that they would take us to the Singleton Hospital and prepared us for the drive. On the way, I was put on the stretcher bed and could see out of the windows the bush rushing by us. I could see the mountains and wondered if the one I saw was the one we had climbed.

Another tidal wave of emotion overcame me, and again the paramedics comforted me with kind and compassionate words. I felt the ambulance come to a stop and saw a council worker holding a sign yell out to the paramedics, "Have you found them yet?"

It surprised me that he knew we were missing. The paramedic driving said, "Yes! We have them in the back now and are taking them to the hospital for a check-over." I could tell the paramedics were happy, and the council worker was too.

"Oh wow, and they both are OK?" The paramedic said.

"Yes. Apart from a few scratches and bites, dehydration, and a lack of food, they are both in reasonably good health considering," he replied cheerfully.

"That's some good news! I'm glad they have been found," the worker sang out before we drove off again.

The paramedics explained to us that every day they would pass these road workers on the way to the national park, and every day, they kept asking if we had been found yet.

Their concern for us made me cry again; it was genuinely heart-warming. I laid down for the drive with my eyes open. If I closed them, I found I could only see us back in the bush.

We arrived in Singleton. It felt so strange being in town again. As we approached the main street of the bridge, the paramedics explained how the police, ambulance, and other specialists had been busy not only looking for us, but also managing an incident where a truck driver had driven into the pub on the corner on one of the days we were lost. My heart felt overwhelmed with sadness thinking about the damage, shock, and trauma experienced by the people who had been affected. "Those poor people," I said.

Just before we arrived at the hospital, the paramedics forewarned me there would possibly be media onsite wanting to capture the first photos of us and that they would try to speak to us and ask lots of questions. I couldn't grasp the concept; I had no idea the extent people had gone–or how many people were aware of us missing or were eager to know how we were or how we had survived.

Arriving at the hospital, the ambulance stopped, and we were prepared to move from the vehicle to the emergency entrance. With me already on a stretcher, they had organised a wheelchair for Dylan en route. The paramedic driver got out and opened the back doors of the ambulance. Immediately, I saw a reporter with a microphone, only later learning it was a Channel 7 reporter named Robert Ovadia, moving as close as he could to us. The paramedic assisted Dylan into the awaiting wheelchair that a nurse was holding onto.

I was halfway out of the ambulance when the electronic bed machine stopped, which had both paramedics puzzled. Within seconds, they realised the batteries had gone flat, so someone quickly replaced them to allow the bed to move the entire way out of the vehicle so the legs could be put down. As they removed the bed and began wheeling me into the hospital, a reporter held his microphone up to his mouth and asked, "How are you feeling?" before putting the microphone to my mouth. "Good," I mumbled, trying to figure out who he was.

When we entered the hospital, natural human curiosity got the better of the medical staff, and we were surrounded by people asking everything from how we were to how we survived. How many days had we been out there, and did we have any food or water with us?

It was a lot to take in. Questions and information bounced back and forth before I could ask if any of our family had been contacted. A nurse beside my bed said the police would have taken care of that. She explained we had been all over the news, and many people had been incredibly worried, adding that it was great to see us both alive and well. "It's a miracle," she told me, "And to see you both alive and well is a miracle too." I didn't know what to say, so I smiled warmly in response.

We were taken from the emergency area into a room with a bed and a door leading to a bathroom with a shower, toilet, and sink. A doctor entered the room, introduced himself, and asked us dozens of questions about our health and what food or drink we had consumed out in the bush, and many of the same questions the emergency teams had asked us.

I felt emotionally and physically exhausted and drained. I just wanted to rest. It was hard to comprehend everything or even think about their questions, let alone answer them all in my state of mind.

I stayed on the ambulance bed in the room while Dylan remained in his wheelchair. I still kept dropping my bundle and had several more emotional bursts that I tried desperately to contain. I looked over at Dylan and closed my eyes, again thanking all the higher powers that he was here, safe, and looking happy being attended to and having a range of food and drinks brought to him. He was like a little rockstar, and I was so proud and overjoyed watching him lapping it all up.

There was so much food made available to us; it was like we were in a hotel with room service. I still did not want or have the taste for food. No matter how much water I drank, I felt constantly thirsty.

A nurse took some blood from me for testing, and as I watched her leave, my mum walked into the room. I burst into tears of joy at the sight of her and recalled the moment I thought I'd never get to see her again. I was so thankful I had survived to feel her arms comforting me again. She cried the instant she saw Dylan and I, too. We put our arms around one another, holding the other tightly like we never wanted to let go of each other again.

After we got through the tears and emotional reunion, we could speak, and Mum's first words were, "Are you both OK?"

I assured her we were okay, but with the doctors and nursing staff rushing in and out like ping pong balls, poking, prodding, and fussing over us, she was really worried. It was a relief when the doctor in charge came in and said he was confident we would only be staying overnight to keep an eye on us so we could expect to go home the following day.

I asked Mum, "Does Daniel, Sarah, and Tim know we are here?"

She nodded and said, "Yes. I contacted everyone. They're on their way."

I leaned back, closing my eyes, grateful that I would get to see my other children soon. "How did you know we were here, Mum?"

She went on to explain how the detective had rung her after the police officers had notified him at the rescue site, and as she recounted, she began to cry.

"Poor Mum," I thought, "What a horrible thing for her to have to go through, and Dad too." He was the quiet type, but he would have been internally worried as well. It would have been heartbreaking for them, the kind I didn't ever want to imagine.

We were asked if there was anything we would like, and I was quick to say, "Yes! A shower!" Dylan was on the same page. He'd had plenty of food and juice and looked forward to washing away the bush.

A nurse came in to give us both a gown, a fresh towel, a toothbrush, and toothpaste. Just feeling them in my hand was the best feeling. I loved brushing my teeth and couldn't wait to remove the furry, dirty feeling from them and wash my body free of the dirt in every crevice I could imagine.

The doctor told us we had to check our bodies for insects or bugs. Puzzled by his request, I looked at my mum; then I began to cry. She read my mind and hugged me, saying, "You're okay now. Dylan is safe. You have both survived." The fear he wouldn't be ok stayed with me. Thinking about how close he was that day to leaving me was just devastating.

A male nurse came in to assist Dylan and help him check his body for any bush bugs or insects. He took the longest shower I had ever known him to have. The nurse checked in on him every 5 minutes to see if he was alright. At one point, he came out and said Dylan had a tick in his groin. My heart raced, but the nurse assured me he was fine, but it would need to be removed, and he would be back. A few minutes later, he returned with the doctor.

Thirty minutes later, he emerged looking refreshed and happy with a smile. "I didn't want to get out," he announced brightly. "The nurse said they found a bug on me," he said.

"Yes, they told me. They are going to remove it," I told him.

Dylan asked, "Are you OK, Mum? Why can't you walk?"

The nurse chimed in and said, "Poor Mum has worked her body so hard to get you both out of the bush. You both need a rest and some sleep in a nice warm bed so her legs can rest."

Dylan walked over to me to hug me before I got ready to go to the bathroom. Tears streamed down my cheeks. I looked at him and said, "I love you so much."

"I love you too, Mum," he answered.

I was so grateful the nurse jumped in and took over me needing to explain it. I was non-stop emotional. I couldn't find my balance, physically or emotionally. With my mum's help, I finished preparing to go to the shower. Dylan was instructed to lie on the bed so they could remove the tick. I asked what type it was, and they told me it was the kind that can make people very sick if the whole bug, head and body, wasn't extracted fully, but they assured me they had done the same procedure too many times to count and not to worry. "Go and have your shower; he's in safe hands."

Mum assisted me from the bed to a different waterproof wheelchair that a nurse wheeled in from the bathroom. "It's fine to get wet so you can remain seated while showering," he told me. Once I was in the chair, Mum wheeled me into the bathroom.

I was so excited and looking forward to the experience of warm water, soap, shampoo, and conditioner. Mum and I joked as she closed the door behind us. It brought back memories of when I was little, and she had to shower me. I said, "It's a bit much at my age that I can't shower myself, isn't it?" and she replied, "It's extreme to get lost, don't you think if all you wanted was some attention from Mum," and we laughed. Humour was our way of coping through the hard times and celebrating the good times.

Remaining sitting in the chair, I turned on the warm water and felt it wash over my bare face and body. I literally did not move for several long minutes. Then, I lathered up my body and felt the dirt drain away from my skin. It took a couple of fresh latherings of soap to wash it properly.

Mum put shampoo in my hair and washed it out for me before repeating it again and then adding the conditioner. It was so wonderful and relaxing—just like being at the hairdressers. There was something special about someone washing your hair while you sat and absorbed the feeling of it. Mum left to check in on Dylan, and I sat and let the water cascade over me, revelling in the feeling of being washed and cleaned.

Mum returned and told me that the doctor was able to remove the whole tick, and they gave Dylan another thorough check-over to make sure there wasn't any more. I had begun rewashing my body, and I spotted a tiny piece of dirt that refused to go, no matter how hard I scrubbed away at it. I scrapped it with my fingernail, but it still refused to budge.

"What the hell?" I thought, bending my head down closely to take a good look at it. With my head down virtually in line with my stomach, I saw tiny legs kicking and squirming but no head. It had burrowed into my belly. Mum went out to let the nurse know I also had a tick, while I brushed my hands over my body to feel for anymore. It was hard to check sitting down.

Cleaning my teeth and brushing my hair, I felt human again. They both felt so wonderful. What a luxury! All these small things we take for granted in our everyday life. Even sitting on a toilet and using toilet paper was so gratifying.

The nurse conveyed the information that I had a tick, and I could hear Mum talking. Seeing this little bug squirming around about half in and half out of my body freaked me out. Mum returned and helped me dry and dress in the gown. It felt like heaven being clean and in clean clothing.

Returning to the room, I saw Dylan perched up in bed, wrapped in his gown with a massive smile, eating custard. He was in his glory. "He will sleep well tonight in a warm bed with blankets and a pillow," I thought, smiling back at him.

The doctor returned to the room with another nurse as Mum was helping me back onto the bed. The doctor explained they needed to remove the tick, and a female nurse would thoroughly check my skin to ensure there were no other wiggly invaders.

I watched the doctor and nurse lean over me and begin pulling the tick out of my stomach with silver tweezers. I laid back while they were having their discussions and put my arm over my eyes, feeling the wave of emotion come through me again as our ordeal played out in my mind.

Dylan was taken for a walk to see the room we would be going into shortly while the doctor worked at extracting my tick. "You wouldn't think something no bigger than an ant would cause so much trouble," I thought. Before long, they had pulled it out and held it in front of me to see before placing it into the specimen jar, where it squirmed around some more, agitated at being yanked out of its burrow.

After a complete examination, I was given the all-clear just as Dylan returned excitedly to the room. "You're going to love it, Mum!" he said animatedly.

"Yeah? So you like it too then, hey?" I asked him.

He went on to tell me there were two comfy beds, a TV, and a kitchen. I chuckled at his excitement. "Wow!" I said, almost meeting him at his level of excitement.

I couldn't tell you the time or how much time had lapsed since we arrived at the hospital, but it didn't feel long before my family began arriving.

When Timothy walked into the room, he didn't say a word. Instead, he rushed over to me, crying and hugging me hard. Eventually, he moved to sit on the end of the bed with his hand on my leg. "What happened? Why did you go out there?" he asked. To think he had thought he had lost his mother and little brother really affected me.

I tried to comfort him but was an emotional wreck myself, attempting to tell him how we had gone off the track and become lost. Still crying, he hugged me again and again, repeating, "I am so glad you're OK." And then asking, "Where is Dylan?"

With all the medics and Mum standing around us, he hadn't noticed Dylan sitting in his wheelchair in the corner of the room. He leapt up and rushed over to him, hugging him, and they both cried. Seeing their love spill over and expressed like it was, made me cry even harder than I was already. I was so proud of them both.

Sarah, my beautiful daughter, walked in and had the same reaction, crying and hugging us, greatly relieved to see her mum and little brother alive and safe. Daniel followed shortly after, saying he thought he would never see us again in between crying.

My heart ached to see my children all together. It pained me to remember how I'd never see them again, and now see the worry I had put them all through.

Friends spilled into our room, and it quickly got crowded. The nurses commented on how popular we were and said they'd leave us for a while to prepare our room for the night and give us some space. Everyone had questions and wanted answers on what happened. I was struggling and felt far too emotional to tell all of our story at that point.

Mum sat beside my bed the whole time, holding my hand, giving it squeezes of comfort and support while discussions were being had, and offering me loving hugs when I broke down.

I was still overcome with the guilt that it was my fault we had gone through the experience, had to endure what we did, and nearly died.

After a while, our friends left, and our family stayed on. The nurses returned to say our room was ready and was larger than the one we were currently in, so it could accommodate everyone more comfortably. Dylan stood up confidently, walked over to my bed, and helped me into the wheelchair he had been using when we first arrived. Everyone collected their belongings, and Mum led the way, wheeling me to the new room with everyone walking directly behind us.

A nurse pushed open the door and announced, "This is it! Our family room!" welcoming us. Inside were two beds parallel to each other with a small walk-through gap between them. A kitchenette with a kettle, microwave, bar fridge, some cupboards, and a sink were positioned in the room, and an ensuite could be seen beyond a slightly opened door. A TV sat on the wall facing the beds.

Dylan climbed into his bed excitedly. Two nurses took hold of my arms and pulled me up to stand. Shaking and bobbing up and down on my wobbly legs, the nurses told me to shuffle my legs until I was at the bed, reassuring me they had me and I wouldn't fall.

I sat down, and the nurses reached down, swung my legs onto the bed, and pulled the covers up and over me. "Is there anything you need?" All I wanted was water. I closed my eyes, feeling drained. I heard Mum ask one of the nurses why I was speaking the way I was and listened to them explaining that it was due to extreme dehydration, starvation, and the stress of the experience.

The nurse came over to me and told me the doctor was still waiting on the bloods, but shortly, they would connect the cannula they had inserted earlier to a drip. "Let's face it, you have worked your body bloody hard and need as much hydration as we can give you." I teared up again. She smiled and patted my arm as

two official-looking people walked into the room. A woman who wore a blouse, work dress pants, and a shiny badge on her hip, and a man who wore a men's shirt with a suit jacket. Entering the room, they introduced themselves as detectives and asked if we were able to speak with them. After my family made all the introductions, everyone except Mum left the room to give us time to talk with the detectives. "Everyone is amazed you made it out alive after so many days lost in the bush with no water or food," they began, "So, can you tell us how you got lost?"

I answered their questions as best I could. Sometimes, the detective would repeat a sentence or a word or ask the question in a different context to gain clarification of my answers. With my speech slurring, sometimes Mum had to repeat my words so they could understand my answers.

Satisfied with my answers, they moved on to Dylan to ask him similar questions to mine, but in a less direct and formal way. I didn't think much about their presence or why they felt they needed to ask so many questions at the time.

After the detectives left, the rest of the family returned, this time with my sister, Belinda, who had arrived while we were being questioned. Lots more tears and hugs followed. It was so great to see her.

I looked at my whole family in this one room and felt the immense love that filled it. I was so thankful Dylan and I got the chance to see and feel it again. If I had ever taken them for granted, I definitely would never again.

This whole ordeal would forever change all of us. We would hug longer, be more present, love louder, and make the most of every moment we got together.

I felt my eyes grow heavy, thinking about them all, smiling at their funny stories, and hearing them laugh and cry. Feeling my Mum's hand on mine, I thought how lucky we all were. Now our recovery begins, but that's another story.

Epilogue

No matter how much I explain or give details of what we went through—the trekking, the climbing, the thought processes, it's tough for people to understand or comprehend the extent of our ordeal fully. I hope that in writing this book, a greater understanding of what Dylan and I and our families went through will occur.

In the beginning, I didn't have the emotional or physical capacity to talk about our experience in great depth. It was a struggle and an ordeal in itself. Writing this book enabled me to finally tell what, in many ways, was our untold story.

I have seen shows on TV such as 'I Shouldn't Be Alive' or 'Got Home Alive', and I can relate strongly to the fear and the devastating feeling of thinking, "Is this it?" Especially for people who had their children with them during the experience. It is hard enough to fight for survival for yourself, but it is even more complicated when you must fight for the survival of your child or children as well.

I promised myself that I would fight and do everything I could to at least make sure Dylan made it out alive, and somehow, I did. They say you can never underestimate a parent's willpower to ensure their child's survival. No matter what the statistics or experts claim, I agree. You have no way of knowing your will, strength, or capabilities until your child's life depends on it.

I was utterly clueless about the scale of the measures put into place to try to find-us. None of the scenarios that played out in my head while in the bush came close to the actual reality of what my family, friends, and the search and rescue teams did. It was nothing short of incredible.

Daniel, as it turned out, did stay on longer at his mate's and didn't return home until Tuesday, the second day of us being lost. He promptly called my daughter Sarah, who rang my work friends to find out where I was. Discovering I hadn't been at work or in touch with anyone else, Sarah was the first to report us missing.

Initially, detectives were suspicious of foul play. Daniel, in particular, was in the firing line, subjected to lengthy and frequent questioning, as he had just returned home from Queensland. It wasn't until they searched our home sometime later and found my sheet of paper with the list of national parks I had written that they had any lead of where we could be. We could have been anywhere, doing anything! The Detectives told me that without that paper list, they wouldn't have thought to look for us in the park as it was so out of character for us. How lucky we were, I remember thinking, and still do!

Going off my list, they began tracking us and our movements. A breakthrough came on Monday, a week after we became lost when detectives located a bird-watching camera placed alongside the road leading into the Mount Royal. Using the footage, they were able to confirm that my car had entered the park Monday morning, but upon further review of the footage, they found our car never exited. A team was sent to Mount Royal to find our car, where we left it in the camping spot.

Authorities immediately set up a search and rescue camp, and ground teams got to work scouring the many bush tracks while choppers took to the air.

Everyone was beside themselves with worry for our safety. Sarah continuously travelled back and forth from her home in Parks to our home in Singleton,-liaising with detectives. Mum spoke with everyone else, keeping them in the loop and communicating updates to all our family and friends.

As the days went on, the concern for our safety heightened. Mum was travelling up from Sydney to provide her statement to Detectives when she got the call everyone was hoping for. We had been found and were alive and safe!

The following is the timeline of our ordeal based on the information I have received.

Day 1. Monday—We entered Mount Royal National Park

Day 2. Tuesday—Daniel returned home and called Sarah; we were reported missing

Day 3. Wednesday—Detectives start their investigation

Day 4. Thursday—Investigation continues

Day 5. Friday—Investigation continues

Day 6. Saturday—Investigation continues

Day 7. Sunday—Investigation continues

Day 8. Monday—Detectives locate bird-camera footage of us driving into the park

Day 9. Tuesday—Search and rescue camp is set up, and the search commences at Mount Royal National Park

Day 10. Wednesday

Day 11. Thursday—Freedom!

I spent ten days in the hospital afterwards with Dylan, who didn't leave my side. The media attention got so great that the hospital had to secretly transfer us to another hospital to recover.

I struggled to eat. The very thought of food made me feel ill. My mum had to force-feed me. Even today, I still can't eat some things. It's like my taste buds were transformed entirely out in the bush.

On the other hand, even when full, Dylan ate everything in sight—he still does! A psychological kickback from starving and the fear of it ever happening again. He gained a good 5 kg within the first week in hospital.

It took about–18 months for my legs not to tingle or feel shaky while walking.

Other than being a bit nervy in the dark, we both escaped our ordeal without a physical scar and with surprisingly few emotional ones.

I still carry the guilt of putting Dylan and my family through such a situation, but every day, I'm thankful we have lived to tell the tale.

Today, when I see a butterfly, I take a few moments to watch it flutter around until it's out of sight. I believe it's a reminder and a sign that everything will be OK.

Acknowledgments

To my youngest son, Dylan,

Your kindness, understanding, and encouragement shown in the bush, in life, and throughout the writing of this book as I put our story into words will be remembered always. I am beyond proud of you and love you with all my heart.

To my three other beautiful children, Daniel, Sarah, and Timothy, words can't express how much I love you all. You are the most incredible human beings. I love you all infinitely.

To my incredible Mum (Ma), we are so lucky to have you in our lives. Thank you for all you did when we finally were free of the bush. Your support and Dad's-are forever appreciated. I love you both.

To my wonderful family and friends, we love you so much for being there for us. It means more than you'll ever know.

Thank you to the SES, police officers, and officials involved in our rescue. We greatly appreciate all your efforts.

Thank you to the paramedics who were there for us in our moment of need with such kindness, understanding and care.

We are deeply grateful to all the staff, nurses, and doctors who provided the care and support we needed during our recovery. It meant so much to us.

To the community and everyone who sent well wishes, kept cheering for us to be found, and who supported my family during the search, thank you so much.

And to Bear Grylls, thank you for making your videos. The knowledge we learned through them helped us to survive the wilderness against the odds.

Printed in Dunstable, United Kingdom

72321913R00134